The ESSENTIALS® of
REGISTERED TRADEMARK

PHYSICAL
CHEMISTRY I

Staff of Research and Education Association,
Dr. M. Fogiel, Director

This book covers the usual course outline of Physical Chemistry I. For more advanced topics, see *"THE ESSENTIALS OF PHYSICAL CHEMISTRY II."*

Research and Education Association
61 Ethel Road West
Piscataway, New Jersey 08854

THE ESSENTIALS ® OF
PHYSICAL CHEMISTRY I

Printed in the United States of America

Library of Congress Catalog Card Number 96-72632

International Standard Book Number 0-87891-620-2

WHAT "THE ESSENTIALS" WILL DO FOR YOU

This book is a review and study guide. It is comprehensive and it is concise.

It helps in preparing for exams, in doing homework, and remains a handy reference source at all times.

It condenses the vast amount of detail characteristic of the subject matter and summarizes the **essentials** of the field.

It will thus save hours of study and preparation time.

The book provides quick access to the important facts, principles, theorems, concepts, and equations of the field.

Materials needed for exams, can be reviewed in summary form — eliminating the need to read and re-read many pages of textbook and class notes. The summaries will even tend to bring detail to mind that had been previously read or noted.

This "ESSENTIALS" book has been carefully prepared by educators and professionals and was subsequently reviewed by another group of editors to assure accuracy and maximum usefulness.

Dr. Max Fogiel
Program Director

CONTENTS

CHAPTER 1

INTRODUCTION

1.1 UNITS

Time, mass and length are the three basic units in science. Newton's Law of Gravitation introduces mass in the equation $F = \dfrac{Gm_1m_2}{d^2}$, where F is the attractive force between two masses m_1 and m_2, which are separated by a distance d. Newton's equation also defines force by the equation $F = ma$.

The second (s) is the basic unit of time. The meter (m) is the basic unit of length, and the kilogram (kg) is the basic unit of mass. Time, mass and length are the base units of the International System of Units (SI units). The prefixes deci-(d), centi-(c), and milli-(m) are used in the SI units for the values 0.1, .01, and .001. The unit of volume is the liter, which is defined as the volume that is occupied by a mass of 1kg of pure water at its maximum density at 4°C and at standard atmospheric pressure (1 atm).

The calorie is the unit of heat and is defined as the amount of heat necessary to raise the temperature of 1g of water 1°C.

1 calorie = 4.184 Joule

1 Volt-coulomb, the energy that a coulomb of electricity gains while passing through a 1 volt potential difference, is known as the joule (J).

1

The watt, Js^{-1}, is the SI unit of power, which is the rate of work. The unit of energy is the watt-hour or kilowatt-hour. The horsepower is 746 watts. The Btu is the unit of heat that is equal to 1055J (or 252 calories).

1.2 MATHEMATICAL REVIEW

$$x^{-a} = \frac{1}{x^a} \qquad (x^a)(x^b) = x^{(a+b)}$$

$$x^a y^a = (xy)^a \qquad \frac{x^a}{x^b} = x^{(a-b)}$$

Suppose M, N and b are positive numbers and $b \neq 1$. If $a = b^d$, then $\log_b(a) = d$, which is the logarithm of a to the base b.

$$\log_b(MN) = \log_b(M) + \log_b(N) \qquad \log_b\left(\frac{M}{N}\right) = \log_b(M) - \log_b(N)$$

$$\log_b(M^P) = p x \log_b(M) \qquad \log_b\left(\frac{1}{M}\right) = -\log_b(M)$$

$$\log_b(b) = 1 \qquad\qquad \log_b(1) = 0$$

The following relationship can be used to change the base of logarithms to c, where $c \neq 1$.

$$\log_b(M) = \log_c(M) \times \log_b(c) = \frac{\log_c(M)}{\log_c(b)}$$

Ordinary logarithms are to the base $b = 10$, and natural logarithms (ln) are to the base $b = e \approx 2.718$.

$$\log_e(x) = \ln(x) = 2.303 \log_{10}(x)$$

$$\log_{10}(x) = .434 \ln(x)$$

The solution for the quadratic equation $ax^2 + bx + c = 0$ is

$$x_{1,2} = \frac{-b \pm \sqrt{b^2 - 4ac}}{2a}$$

2

and its roots are either real or imaginary.

The following are derivatives of some functions:

$$\frac{d}{dx}(uv) = u\frac{dv}{dx} + v\frac{du}{dx} \qquad\qquad \frac{d}{dx}(x^n) = nx^{n-1}$$

$$\frac{d}{dx}(\ln u) = \frac{1}{u}\frac{du}{dx} \qquad\qquad \frac{d}{dx}(e^u) = e^u\frac{du}{dx}$$

An integral is the area under a curve, which is the opposite of a differential.

$$\int dx = x + \text{constant} \qquad\qquad \int_{x_1}^{x_2} dx = x_2 - x_1$$

$$\int x^n dx = \frac{1}{n+1} x^{(n+1)} \qquad \int \frac{dx}{x} = \ln x \qquad \int e^x dx = e^x$$

In the last three equations constants of integration were omitted.

For $u = u(x,y,\ldots,z)$, which is a function of more than one variable, the change of u relative to any one of the variables, while the remaining variables stay constant, is specified by the partial derivative function. The partial derivative of u with respect to x, holding y and z constant, is as follows:

$$\left(\frac{\partial u}{\partial x}\right)_{y,z}$$

To determine the second partial derivative, differentiation is carried out twice.

$$\frac{\partial^2 u}{\partial x \partial y} = \left[\frac{\partial}{\partial x}\left(\frac{\partial u}{\partial y}\right)_x\right]_y \qquad\qquad (1)$$

$$\frac{\partial^2 u}{\partial y \partial x} = \left[\frac{\partial}{\partial y}\left(\frac{\partial u}{\partial x}\right)_y\right]_x \qquad\qquad (2)$$

(1) and (2) are equivalent.

3

For a function of two variables, $u = u(x,y)$, du is the total differential of u and is written in the form

$$du = \left(\frac{\partial u}{\partial x}\right) dx + \left(\frac{\partial u}{\partial y}\right) dy$$

If $M(x,y) = \dfrac{\partial u}{\partial x}$ and $N(x,y) = \dfrac{\partial u}{\partial y}$, then $du = Mdx + Ndy$.

$$\left(\left(\frac{\partial M}{\partial y}\right)_x = \left(\frac{\partial N}{\partial x}\right)_y\right)$$

The above equation is called the Euler condition for exactness and u, is then called a state function.

The integral of a state function u around any closed path (where the initial and final points of the integrating path are the same) should be zero, or

$$\oint du = 0$$

Some useful relations between partial derivatives are:

$$\left(\frac{\partial x}{\partial y}\right)_z = \frac{1}{(\partial y / \partial x)_z}$$

$$\left(\frac{\partial x}{\partial y}\right)_z = \frac{(\partial x / \partial w)_z}{(\partial y / \partial w)_z}$$

$$\left(\frac{\partial x}{\partial y}\right)_z = \frac{-(\partial z / \partial y)_x}{(\partial z / \partial x)_y}$$

The coefficient of thermal expansion (α) is expressed by the equation

$$\alpha = \frac{1}{V}\left(\frac{\partial V}{\partial T}\right)_P$$

The compressibility (β) is expressed by the equation

$$\beta = -\frac{1}{V}\left(\frac{\partial v}{\partial p}\right)_T$$

$$\left(\frac{\partial P}{\partial T}\right)_V = \frac{-(\partial v / \partial T)_P}{(\partial v / \partial p)_T} = \frac{\alpha}{\beta}$$

CHAPTER 2

GASES

2.1 PRESSURE AND TEMPERATURE

PRESSURE

Pressure is force per unit area. The units used to express pressure are shown in Table 2.1.

The absolute pressure of a system is equal to the gauge pressure plus the ambient (surrounding) atmospheric pressure.

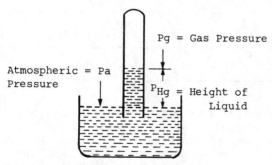

Fig. 2.1 A simple apparatus used to determine the pressure of a liquid.

$$Pa = Pg + PHg$$

$$h_{liq} = h_{Hg} \frac{d_{Hg}}{d_{liq}}$$

where h_{liq} and h_{Hg} are the heights of the liquid and the mercury, respectively, in the column, and d_{liq} and d_{Hg} are the densities of the liquid and the mercury, respectively.

Table 2.1. The units of pressure

Unit	Definition	Equal to	Symbol
Pascal		1 Newton/meter^2 (N/m^2 or $kg/m \cdot s^2$)	Pa
Atmosphere	The force exerted by a column of mercury 760mm high/unit area	$1.01325 \times 10^5 N/m^2$ or $1.01325 \frac{dynes}{cm^2}$ 760 torr, 760 mmHg	Atm
Torr	The pressure exerted by a 1mm column of mercury/unit area	$133.3224 N/m^2$	Torr
Pound inch^2		$6894.7572 N/m^2$	lb/in^2 or
Bar		$1 \times 10^5 N/m^2$	Psi
Micron		1×10^{-6} torr	μtorr

The pressure-measuring devices that are used in laboratories are: the aneroid barometer, the precision-type bellows manometer, the Mcleod gauge, the diaphragm manometer, the capacitance manometer and the ionization gauge. Some of these devices are used to measure low pressures.

TEMPERATURE

On the Celsius or centigrade scale, the fixed value for the freezing point of water at 1 atm is 0 °C and that for the boiling point of water at 1 atm is 100°C.

6

$$T(^\circ K) = T(^\circ C) + 273.15^\circ$$

where $T(^\circ K)$ is the absolute Kelvin scale.

On the Fahrenheit scale, the fixed value for the freezing point of a saturated NaCl-water solution is $0^\circ F$ and that for the boiling point of pure water is $212^\circ F$.

$$T(^\circ F) = \frac{9T}{5}(^\circ C) + 32$$

2.2 LAWS OF IDEAL GASES

A gas that obeys the equation of state $PV = nRT$ is an ideal gas. All gases increasingly obey the ideal gas law at increasingly low pressures; therefore, the equation $PV = nRT$ is a limiting law for the description of real gases.

Molecules of ideal gases occupy no volume and no interaction forces take place among them.

BOYLE'S LAW

Boyle's Law states that at constant temperature the volume of a gas is inversely proportional to the pressure:

$$V\alpha\,\frac{1}{P}, \quad or \quad PV = constant$$

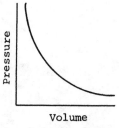
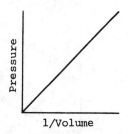

Fig. 2.2 Applying Boyle's law by plotting pressure as a function of volume.

Fig. 2.3 Applying Boyle's law by plotting pressure as a function of the inverse of volume.

$$P_1 V_1 = P_2 V_2 \qquad \text{(at constant temperature)}$$

LAW OF GAY-LUSSAC AND CHARLES

The Law of Gay-Lussac states that at constant pressure, the volume of a gas is proportional to its temperature, and that at constant volume, the pressure is proportional to its temperature:

$$V \propto T \qquad \text{(at constant P)}$$

$$P \propto T \qquad \text{(at constant V)}$$

$$\frac{V}{T} = \text{constant} \qquad \text{(at constant P)}$$

$$\frac{P_2}{T_2} = \frac{P_1}{T_1} \qquad \text{(at constant V)}$$

$$\frac{V_2}{T_2} = \frac{V_1}{T_1} \qquad \text{(at constant P)}$$

$$PV = B(t + 273.15)$$

$$PV = BT$$

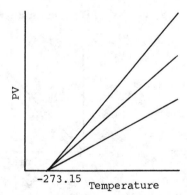

Fig. 2.4 PV as a function of temperature for different quantities of gases.

T is the absolute temperature ($^\circ$K). t is the temperature in degrees Celsius ($^\circ$C).

AVOGADRO'S LAW

Avogadro's Law states that equal volumes of ideal gases contain equal numbers of molecules under the same temperature and pressure conditions. The molar volume (Vm) is the volume occupied by 1 mole of a gas.

COMBINED GAS LAW

$$\frac{PV}{T} = \text{constant}$$

$$\frac{P_1 V_1}{T_1} = \frac{P_2 V_2}{T_2}$$

$$PV = nRT$$

where n is the number of moles of gas, and R is the gas constant.

DALTON'S LAW OF PARTIAL PRESSURES

Dalton's law of partial pressures states that the total pressure of an ideal gas mixture is equal to the sum of the pressures exerted by the individual gases (in the mixture) if placed in the same vessel alone.

$$P_{tot} = P_1 + P_2 + P_3 + \ldots + P_n$$

$$P_{tot} = \sum_j P_j$$

where P_j is the partial pressure of component j in the mixture.

$$P_j = X_j P_{tot}$$

where X_j is the mole fraction of component j.

$$P_j = \left(\frac{n_j}{n_{total}} \right) \times P_{tot}$$

$$X's$$

9

$$P_{tot} = n_1 \frac{RT}{V} + n_2 \frac{RT}{V} + n_3 \frac{RT}{V} + \ldots + n_n \frac{RT}{V}$$

$$P_{tot} = \frac{RT}{V} \sum_j n_j$$

$$X_j = \frac{P_i}{P_{tot}} = \frac{n_i(RT/V)}{\sum_j n_j(RT/V)} = \frac{n_i}{\sum_j n_j} = \frac{n_i}{n_{total}}$$

MOLECULAR WEIGHT OF AN IDEAL GAS

$$M = \frac{mRT}{PV}$$

where m is the mass, and M is the molecular weight.

$$M = \frac{dRT}{P}$$

where d is the desnity.

2.3 REAL GASES

The deviations of real gases from ideality are caused by interactions between individual molecules.

The compressibility factor $Z = P\bar{V}/RT$ measures the deviation from ideality. The amount that Z deviates from unity measures the lack of ideality in a real or imperfect or non-ideal gas.

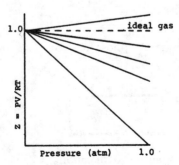

Fig. 2.5 A plot of Z against pressure at low pressures.

The excluded volume ($V_{excluded}$) is the volume that the gas molecules occupy because of their finite size. As the pressure increases, the relative error in neglecting $V_{excluded}$ increases.

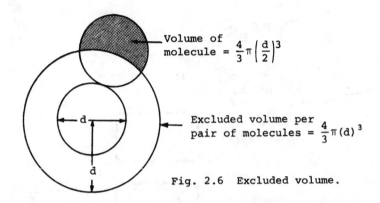

Volume of molecule = $\frac{4}{3}\pi\left(\frac{d}{2}\right)^3$

Excluded volume per pair of molecules = $\frac{4}{3}\pi(d)^3$

Fig. 2.6 Excluded volume.

THE VAN DER WAALS EQUATION

The Van der Waals equation of state is

$$\left(P + a\frac{n^2}{V^2}\right)(V - nb) = nRT$$

where b is the excluded volume of the molecules, and 'a' is a constant. an^2/V^2 is often called the internal pressure of the gas.

The critical point is the point of inflection. At this point the temperature is called the critical temperature T_c, the pressure is called the critical pressure, P_c, and the volume is called the critical volume V_c. The gas phase and the liquid phase are continuous at the critical temperature. A liquid will not form above the critical temperature, regardless of how great the pressure is.

$$P_c = \frac{RT_c}{V_{m,c} - b} - \frac{a}{V_{m,c}^2}$$

11

$$T_c = \frac{8a}{27bR} \qquad V_{m,c} = 3b \qquad P_c = \frac{a}{27b^2}$$

$$b = \frac{V_{m,c}}{3} \qquad a = 3P_c V_{m,c}^2 \qquad R = \frac{8P_c V_{m,c}}{3T_c}$$

$$Z_c = \frac{P_c V_{m,c}}{RT_c} = \frac{3}{8} = .375$$

The reduced variables are the ratios of the actual variables to the critical variables:

reduced pressure:
$$P_R = \frac{P}{P_c}$$

reduced temperature:
$$T_R = \frac{T}{T_c}$$

reduced volume:
$$V_R = \frac{V_m}{V_{m,c}}$$

$$P_r = \frac{8T_r}{3V_r - 1} - \frac{3}{V_r^2}$$

The law of corresponding states proposes that real gases with the same values for any two of the reduced variables will have the same values for the third variable.

The value of z is determined from the graph of z_r plotted against P_r for different reduced temperatures.

MOLECULAR WEIGHT OF A REAL GAS

For n moles of a real gas at low pressures

$$PV = n(RT + B_p P)$$

$$n = \frac{m}{M} \quad \text{and} \quad d = \frac{m}{V}$$

12

$$\frac{d}{P} = \frac{M/RT}{1 + (B_p P/RT)}$$

At low pressures, $[1 + (B_p P/RT)]^{-1} \backsimeq 1 - (B_p P/RT)$ and

$$\frac{d}{P} = \frac{M}{RT} + \left(\frac{M}{RT}\right)\left(\frac{-B_p}{RT}\right) P$$

$\left(\dfrac{M}{RT}\right)\left(\dfrac{-B_p}{RT}\right)$ is the slope of the isothermal plot of d/P as a function of P, and $\dfrac{M}{RT}$ is the intercept of the same plot.

OTHER EQUATIONS OF STATE

The virial equation

$$\frac{PV}{nRT} = 1 + B(T)\left(\frac{n}{V}\right) + C(T)\left(\frac{n}{V}\right)^2 + D(T)\left(\frac{n}{V}\right)^3 + \dots$$

B(T) is the second virial coefficient, C(T) is the third virial coefficient, and so on.

$$B(T) = 2\pi N_o \int_o^\infty \left[1 - \exp\left(\frac{-V_{(r)}}{kT}\right) \right] r^2 dr$$

where N_o is Avogadro's number, k is the Boltzmann constant, r is the distance between molecules, and $V_{(r)}$ is the intermolecular pair potential.

THE BERTHELOT EQUATION

$$P = \frac{nRT}{(V_{m,c} - nb)} - \frac{n^2 a}{T V_{m,c}^2}$$

$$P_c = \frac{1}{12}\left(\frac{2aR}{3b^3}\right)^{\frac{1}{2}} \qquad V_{m,c} = 3b \qquad T_c = \frac{2}{3}\left(\frac{2a}{3bR}\right)^{\frac{1}{2}}$$

$$Z_c = \frac{3}{8} = .375$$

$$P_r = \frac{8T_r}{(3V_r - 1)} - \frac{3}{T_r V_r^2}$$

13

THE DIETERICI EQUATION

$$P = \left(\frac{nRT}{(V_{m,c} - nb)}\right) \exp\left(\frac{-na}{V_{m,c}RT}\right)$$

$$P_c = \frac{a}{4e^2b^2} \qquad V_{m,c} = 2b \qquad T_c = \frac{a}{4bR}$$

$$Z_c = \frac{2}{e^2} = .2706$$

$$P_r = \left\{\frac{e^2 T_r}{(2V_r - 1)}\right\} \exp\left(\frac{-2}{T_r V_r}\right)$$

(e is the exponential)

THE BEATTIE-BRIDGEMAN EQUATION

$$P = \frac{n^2 RT}{V^2}(1 - \gamma)\left[\frac{V}{n} + \beta\right] - \frac{\alpha n^2}{V^2}$$

$$\alpha = a_o\left[1 + \frac{an}{V}\right]$$

$$\beta = b_o\left[1 - \frac{bn}{V}\right]$$

$$\gamma = \frac{nc_o}{VT^3}$$

CHAPTER 3

UNIV.

Sys.

SURROUNDING

THE FIRST LAW OF THERMODYNAMICS

THERM. ENERGY IS AN EXTENSIVE PROPERTY SINCE IT DEPENDS UPON THE AMOUNT OF MATERIAL PRESENT, WHILEAS, TEMP. IS INTENSIVE PROPERTY SINCE IT IS ASSOCIATED WITH THE MOTION OF MOLECULES, AND DOESN'T DEPEND UPON THE AMOUNT OF MAT. PRESENT

3.1 HEAT

Heat is energy in transit. Heat flows from a hot medium to a cold medium. Thermal energy is the random motion stimulated by heat. Thermal energy is an extensive property because it depends upon the amount of material present. Temperature is related to the thermal energy, since it is associated with the motion of molecules. Temperature is an intensive property because its value does not depend upon the amount of material present.

The universe is composed of a system and its surroundings. A reaction vessel, an engine, or an electric cell is an example of the system. The rest of the universe is called the surroundings.

The universe is in thermal equilibrium when the temperature of the surroundings does not change as the experiment proceeds. As a result, there will be no tendency for the heat to flow between a system and its surroundings.

In closed systems, the transfer of matter to and from the surroundings is not permitted. An isolated system is a closed system that does not exchange heat with its surroundings.

The process is isothermal when constant temperature is maintained by heat flow. The process is adiabatic when no heat enters or leaves the system. During this process the

15

(ISOL (+∞))
(CLOSED Sys.)

EXoTHERMIC → HEAT IS LIBARATED FRom THE
System to Surround.

temperature of the system changes. When heat flows from a system to the surroundings, the heat is negative, and the process is exothermic. When heat flows from the surroundings to a system, the heat is positive, and the process is endothermic.

ENDoTHERMIC → HEAT IS ConSUMED By THE
Syst FRom THE ScrRounrmg

3.2 WORK

Work is a way to transfer energy through the system's walls. Mechanical work is done when an applied force $F(x)$ moves an object a distance x.

$$dw = Fdx$$

The force is constant in direction and magnitude.

$$w = \int_{x_1}^{x_2} Fdx = F(x_2 - x_1)$$

where x_2 is the final position, and x_1 is the initial position.

The stretching of a spring is an example of a nonconstant force. Hooke's law expresses the relationship between the force and the amount of stretching.

$$F = -kx$$

where k is the spring constant, and x is the displacement from the equilibrium position of the spring.

$$dw = -kxdx$$

The work done in stretching the spring from its equilibrium position (x = 0) to some position x is

$$w = -\int_{x=0}^{x} kxdx = -\tfrac{1}{2}kx^2$$

Work done on the system is positive and work done by the system is negative.

EXPANSION WORK

Work done on the system when the gas volume changes by dV against an external pressure P_{ext} is

$$dw = -P_{ext}dV$$

$$w_{exp} = - \int_{V_1}^{V_2} P_{ext}dV = -P_{ext}(V_2 - V_1)$$

where V_2 is the final volume, and V_1 is the initial volume.

For a free expansion into a vacuum $P_{ext} = 0$ and $dw = 0$.

ISOTHERMAL REVERSIBLE EXPANSION WORK OF IDEAL GASES

A reversible process goes through an infinite series of equilibrium states as it proceeds from the initial to the final state. In a reversible process, the internal pressure is always equal to the external, opposing pressure.

$$dw = -P_{ext}dV = -P_{int}dV$$

$$w_{exp} = - \int_{V_1}^{V_2} P_{ext}dV = - \int_{V_1}^{V_2} nRT \frac{dV}{V}$$

$$w_{exp} = -nRT \ln\left(\frac{V_2}{V_1}\right)$$

where T is constant.

Since $P_1V_1 = P_2V_2$ for an ideal gas,

$$w_{exp} = -nRT \ln\left(\frac{V_2}{V_1}\right) = -nRT \ln\left(\frac{P_1}{P_2}\right)$$

ISOTHERMAL IRREVERSIBLE EXPANSION WORK OF IDEAL GASES

An irreversible process is not reversible. In an irreversible process, the internal pressure is not always equal to the external pressure.

$$-w = P_2(V_2 - V_1) = P_2\left(\frac{nRT}{P_2} - \frac{nRT}{P_1}\right)$$

17

$$-w = nRT \left(1 - \frac{P_2}{P_1}\right)$$

where P_2 is the final pressure, P_1 is the initial pressure, V_2 is the final volume, and V_1 is the initial volume.

A reversible process produces more work than an irreversible process with the same initial and final states as the reversible process. Reversible work is the maximum amount of work. The closer the system is to equilibrium, the more work is obtained.

WORK AND P-V DIAGRAMS: CYCLIC PROCESS

A cyclic process is the process that ends at its starting point.

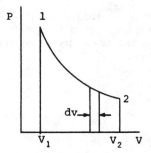

Fig. 3.1 Expansion work for a single-stage process.

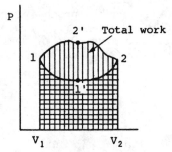

Fig. 3.2 Expansion work for a cyclic process.

The total work in Figure 3.1 is

$$\left(- \int_1^2 PdV \right) + \left(- \int_2^1 PdV \right)$$

$$= - \int_1^2 PdV + \int_1^2 PdV = 0.$$

The total work in Figure 3.2 is the area enclosed by the curve 11'22'.

Work is not a state function because it depends upon the

18

path taken and not simply the state of the system.

The various types of work in thermodynamics are listed below:

mechanical work	$dw = fdx$
surface work	$dw = \gamma dA$
electrical work	$dw = \varepsilon dq$
gravitational work	$dw = mgdx$
expansion work	$dw = -PdV$

where f is force, x is distance, γ is surface tension, A is surface area, ε is potential difference, q is charge (current times time: $dq = Idt$), m is mass, g is gravitational acceleration, P is the pressure exerted on a system by the surroundings, and V is the volume of the system.

3.3 THE FIRST LAW
OF THERMODYNAMICS

The First Law of Thermodynamics states that the total energy of the system plus that of the surroundings is constant.

$$\Delta E_{tot} = \Delta E_{syst} + \Delta E_{surr} = 0$$

$$\Delta E_{syst} = -\Delta E_{surr}$$

where E is the internal energy.

$$\Delta E = E_{final} - E_{initial}$$

$$\boxed{\Delta E = q + w}$$

where q is heat added to the system and w is work done on the system.

CHAPTER 4

APPLYING THE FIRST
LAW OF THERMODYNAMICS

4.1 MANIPULATING THE FIRST LAW

ΔE as constant-volume heat

$$\Delta E = q + w$$

$$dE = dq + dw_e - P_{op} dV$$

where P_{op} is external pressure and w_e is external non-P-V work. $dV = 0$ and $dw_e = 0$ for a constant-volume process.

$$dE = dq_V$$

where q_V is heat absorbed at constant volume.

$$\Delta E = q_V$$

E is a state function because it depends only on the present state of the system and not on the path by which this state was achieved.

A bomb calorimeter is used to measure the heat of combustion for a substance at a constant volume, which is then equal to the change in internal energy (ΔE).

Enthalpy as constant-pressure heat

$$\Delta E = E_2 - E_1 = q_p + w$$

where q_p is constant-pressure heat.

$$H = E + PV$$

where H is the enthalpy.

$$\Delta H = \Delta E + P \Delta V \qquad \text{(at constant pressure)}$$

$$\Delta H = H_2 - H_1 = q_p$$

ΔH is a state function since it depends only on the initial and final states of the process and not on the path of the process.

4.2 HEAT CAPACITIES

Heat capacities are a measure of the heat required to change the temperature of a substance. Heat capacity is an extensive property since it depends upon the amount of material present. The molar heat capacity, C_m, is the heat capacity of 1 mole of material, and it is related to the specific heat, c, by the equation

$$C_m = M_c$$

where M is the molecular weight of the material. The heat capacity is also dependent upon the conditions under which the heat is transferred into the system.

$$C_v = \frac{dq_v}{dT} = \left(\frac{dE}{dT} \right)_v \qquad \text{(at constant volume)}$$

$$C_p = \frac{dq_p}{dT} = \left(\frac{dH}{dT} \right)_p \qquad \text{(at constant pressure)}$$

where C_v is the heat capacity at constant volume, and C_p is the heat capacity at constant pressure.

$C_p > C_v$ because it takes more heat per unit temperature rise at constant pressure than at constant volume. The difference between C_p and C_v is much greater for gases than it is for liquids and solids.

$$C_p - C_v = \left[P + \left(\frac{\partial E}{\partial V} \right)_T \right] \left(\frac{\partial V}{\partial T} \right)_P$$

where $\left(\frac{\partial E}{\partial V} \right)_T$ is the internal pressure.

For ideal gases $V = \frac{nRT}{P}$ and $\left(\frac{\partial E}{\partial V} \right)_T = 0$, and

$$C_p - C_v = nR$$

$$C_p - C_v = \left(\frac{\alpha^2}{\beta} \right) TV$$

where α is the isobaric coefficient of thermal expansion, and β is the isothermal compressibility factor. They are given by the following equations:

$$\alpha = \frac{1}{V} \left(\frac{\partial V}{\partial T} \right)_P = \left(\frac{\partial \ln V}{\partial T} \right)_P$$

$$\beta = -\frac{1}{V} \left(\frac{\partial V}{\partial P} \right)_T = -\left(\frac{\partial \ln V}{\partial P} \right)_T$$

Heat capacities of real substances are dependent upon temperature and are given by the equations

$$C_p = a + bT + cT^2 + dT^3$$

$$C_v = a + bT + c^1 T^{-2}$$

where a, b, c, c^1, and d are constants for each substance.

4.3 JOULE-THOMSON EXPANSION

Joule-Thomson expansion is a constant-enthalpy process. The Joule-Thomson expansion coefficient is defined as

$$\mu = \left(\frac{\partial T}{\partial P}\right)_H$$

When μ is positive, the gas cools on expansion. When μ is negative, the gas warms on expansion. μ is negative above the inversion temperature. The inversion temperature is the temperature at which the sign changes. Most gases have positive Joule-Thomson coefficients and they cool on expansion at room temperature. $\mu = 0$ for an ideal gas, and its temperature does not change during a Joule-Thomson expansion. The Joule-Thomson expansion is important in the liquefaction of gases.

$$\left(\frac{\partial H}{\partial P}\right)_T = -\mu_{jt} C_p$$

$$\left(\frac{\partial H}{\partial T}\right)_V = \left(1 - \frac{\alpha}{\beta} \mu_{jt}\right) C_p$$

4.4 ISOTHERMAL AND ADIABATIC; REVERSIBLE AND IRREVERSIBLE PROCESSES

Isothermal expansion of an ideal gas

$$\Delta E = 0$$

$$\Delta E = q + w = 0$$

$$q = -w$$

Isothermal, isobaric (constant pressure) expansion of an ideal gas

$$\Delta E = \Delta H = 0$$

$$q = -w = -\int_{V_1}^{V_2} PdV = -P\Delta V$$

Isothermal, isobaric phase change

$$q = \Delta H$$

$$-w = \int_{V_1}^{V_2} PdV = P\Delta V$$

$$\Delta E = q + w$$

Isothermal, reversible expansion of an ideal gas

$$\Delta E = 0$$

$$\Delta H = \Delta E + \Delta(PV) = 0 + \Delta(nRT) = 0$$

$$q = -w = nRT \ln\left(\frac{V_2}{V_1}\right) = nRT \ln\left(\frac{P_1}{P_2}\right)$$

Isothermal, irreversible expansion of an ideal gas

$$q = -w = \int_{V_1}^{V_2} P_{op}dV = P_2 \int_{V_1}^{V_2} dV = p_2(V_2 - V_1)$$

$$= P_2\left(\frac{nRT}{P_2} - \frac{nRT}{P_1}\right)$$

$$q = -w = nRT\left(1 - \frac{P_2}{P_1}\right) = nRT\left(1 - \frac{V_1}{V_2}\right)$$

An isothermal, irreversible compression of constant pressure is impossible.

Adiabatic expansion of a gas

$$q = 0 \quad \text{(For an adiabatic expansion)}$$

$$\Delta E = q + w = o + w$$

$$dE = nCvdT = dw = -PdV$$

where n is the number of moles of gas.

Reversible, adiabatic expansion of an ideal gas

$$q = 0$$

$$\Delta E = +w = \int_{T_1}^{T_2} nC_v dT$$

$$\Delta H = \int_{T_1}^{T_2} nC_p dT$$

$$(T_1)^{\frac{C_v}{nR}} V = (T_2)^{\frac{C_v}{nR}} V_2$$

$$\frac{T_1}{T_2} = \left(\frac{V_2}{V_1}\right)^{(\gamma - 1)}$$

where $\quad \gamma = \dfrac{C_p}{C_v} \quad$ and $\quad C_p - C_v = nR$

$$P_1 V_1^{\gamma} = P_2 V_2^{\gamma}$$

$$PV^{\gamma} = \text{constant}$$

The gas is cooled during an adiabatic expansion. In an adiabatic compression, however, the signs for all the quantities are reversed; work is done on the gas and the temperature of the gas increases.

Isobaric, adiabatic expansion of an ideal gas

$$q = 0$$

$$w = - \int_{V_1}^{V_2} PdV = -P \Delta V$$

$$\Delta E = \int_{T_1}^{T_2} nC_v dT$$

$$\Delta H = \int_{T_1}^{T_2} nC_p dT$$

$$\int_{T_1}^{T_2} nC_v dT = -P \Delta V$$

Irreversible, adiabatic expansion of an ideal gas

$$nC_v(T_2 - T_1) = -P_2(V_2 - V_1) = -P_2 \left(\frac{nRT_2}{P_2} - \frac{nRT_1}{P_1} \right)$$

$$nC_v(T_2 - T_1) = -nR \left(T_2 - \frac{T_1 P_2}{P_1} \right)$$

$$C_v(T_2 - T_1) = -R \left(T_2 - \frac{T_1 P_2}{P_1} \right)$$

CHAPTER 5

THERMOCHEMISTRY

5.1 HEATS OF REACTION

$$\Delta H = \Delta E + \Delta n \text{ gas } RT \quad \text{(for a reaction)}$$

$$\Delta n \text{ gas} = n \text{ gas(products)} - n \text{ gas(reactants)}$$

where n gas is the total number of moles of gas present as products or as reactants.

The heat of formation is the heat of the reaction for the production of a compound from its elements. ΔH_T^o (formation) is the heat of formation at the standard state of 1 atm and 298.15°K. The heat of formation for an element is zero in its standard state.

$$\Delta H_T^o(\text{reaction}) = \sum_i^{\text{products}} n_i \Delta H_T^o(\text{formation, i})$$

$$- \sum_j^{\text{reactants}} n_j \Delta H_T^o(\text{formation, j})$$

where n_i and n_j are the number of moles o products and reactants, respectively.

HESS'S LAW

Hess's law states that if a series of reactions having known ΔH values can be arranged in such a way that, when added up, they produce a desired equation, the ΔH value for the desired reaction will be equal to the sum of the ΔH values for the individual reactions.

$$\Delta H_{tot} = \Delta H_1 + \Delta H_2 + \ldots$$

Heat of combustion is the heat of the reaction for the total oxidation of one mole of a compound.

$$\Delta H^o_T(\text{reaction}) = \sum_i^{\text{reactants}} n_i \Delta H^o_T(\text{combustion, i})$$

$$- \sum_j^{\text{products}} n_j \Delta H^o_T(\text{combustion, j})$$

Heat of neutralization is the heat of the reaction for the neutralization of an acid by a base.

The (integral) heat of solution is the heat of the reaction for dissolving a mole of solute in n moles of solvent. ΔH^o (solution) for gaseous solute results from the solvation of the solute molecules. The solution process for a solid molecular solute is the sum of two processes:

solute (s) = solute (molecules) $\qquad\qquad \Delta H^o_T(\text{sublimation})$

solute(molecules) + solvent = solution $\qquad \Delta H^o_T(\text{solvation})$

solute (s) + solvent = solution

$$\Delta H^o_T(\text{solution}) = \Delta H^o_T(\text{sublimation}) + \Delta H^o_T(\text{solvation})$$

The integral heat of solution at infinite dilution is the heat of the reaction for dissolving a substance in an infinite amount of solvent.

The (integral) heat of dilution is the heat of the reaction for diluting one mole of a solute in a solution of given concentration, by adding solvent to produce a solution of different concentration.

The bond (dissociation) energy (or enthalpy) is the heat of the reaction that determines the strengths of bonds. It results from the breaking of chemical bonds in gaseous molecules.

When bond energies are used to calculate the heats of reaction, it is assumed that the following steps take place in the reaction: (1) decomposition of the reactants into molecular fragments and (2) formation of the products from the fragments.

28

$$\Delta H^o_T \text{(reaction)} = \sum_i^{\text{reactants}} n_i BE_i - \sum_j^{\text{products}} n_j BE_j$$

where BE is the bond energy and n_i and n_j are the number of moles of bonds involved in the reaction.

The enthalpy of atomization is the heat of the reaction required to break a molecule into its component atoms. Enthalpy of sublimation is a special case of the enthalpy of atomization and also a special case of the enthalpy of phase transition. Evaporation, melting and changes of crystal form are examples of phase transitions.

5.2 TEMPERATURE DEPENDENCE OF REACTION ENTHALPIES

$$\Delta H_T = \Delta H_{298^o} + \int_{298}^{T} \Delta C_p \, dT$$

$$\Delta C_p = \sum_{\text{products}} nCp - \sum_{\text{reactants}} nCp$$

If $Cp = a + bT + cT^2 + dT^3$

$$\Delta H_T = \Delta H_{298} + \Delta a(T - 298) + \frac{\Delta b}{2}(T^2 - 298^2) + \frac{\Delta c}{3}(T^3 - 298^3)$$

$$+ \frac{\Delta d}{4}(T^4 - 298^4)$$

$$\Delta H_T = \Delta H_0 + \Delta aT + \frac{\Delta b}{2}(T^2) + \frac{\Delta c}{3}(T^3) + \frac{\Delta d}{4}(T^4)$$

where

$$\Delta H_0 = \Delta H_{298} - (\Delta a)(298) - \frac{\Delta b}{2}(298^2) - \frac{1}{3}(\Delta c)(298)^3$$

$$- \frac{1}{4}(\Delta d)(298)^4$$

and $\quad \Delta a = \Sigma \underset{\text{products}}{na} - \Sigma \underset{\text{reactants}}{na}$

If $Cp = a + bT + cT^{-2}$

$$\Delta H_T = \Delta H_0 + \Delta aT + \frac{\Delta b}{2} T^2 - \Delta c(T)^{-1}$$

where

$$\Delta H_0 = \Delta H_{298} - \Delta a(298) - \frac{\Delta b}{2}(298)^2 + \Delta c(298)^{-1}$$

$$\Delta H_{T_2} = \Delta H_{T_1} + \int_{T_1}^{T_2} \Delta Cp\, dT$$

5.3 HEAT AND PHYSICAL CHANGES

For each phase transition, there is an energy change, e.g.

$S \rightarrow liq$, $\quad \Delta H_T^o$(fusion); $\quad liq \rightarrow S$, ΔH_T^o(crystallization);

$S \rightarrow gas$, $\quad \Delta H_T^o$(sublimation); $\quad liq \rightarrow gas$, ΔH_T^o(vaporization)

ΔH_T^o(fusion) $= - \Delta H_T^o$(crystallization)

Approximation of the heats of transition

For substances that are not highly joined in the liquid state, the following relationships have been noticed:

ΔH_T^o(vaporization) $\cong$ (88J k^{-1} mol^{-1})T_{bp}

where T_{bp} is the normal boiling point of the liquid. For elements,

ΔH_T^o(fusion) $\cong$ (9.2J k^{-1} mol^{-1})T_{mp}

where T_{mp} is the melting point of the solid.

ENTHALPY OF HEATING

$$\Delta H^\circ = \overset{phases}{\underset{i}{\Sigma}} \int_i Cp^\circ \, dT + \overset{transitions}{\underset{j}{\Sigma}} \Delta H^\circ_T (\text{transition}, j)$$

The following example shows how the above equation is applied.

Example:

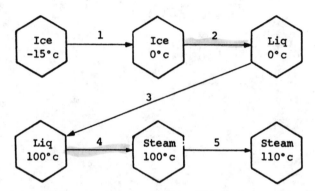

Fig. 5.1 Heating ice from -15°c to steam at 110°c.

$$\Delta H^\circ = \Delta H^\circ_{(1)} + \Delta H^\circ_{(2)} + \Delta H^\circ_{(3)} + \Delta H^\circ_{(4)} + \Delta H^\circ_{(5)}$$

$$= \int_{258k}^{273k} Cp^\circ dT + \Delta H^\circ_{273} (\text{fusion}) + \int_{273k}^{373k} Cp^\circ dT$$

$$+ \Delta H^\circ_{373} (\text{vaporization}) + \int_{373k}^{383k} Cp^\circ \, dt$$

CLAPEYRON EQUATION

When the phase changes involve only condensed phases, the following Clapeyron equation must be used.

$$\frac{dp}{dT} = \frac{\Delta H}{T \Delta V}$$

where ΔH is the enthalpy change for the transition and ΔV is the corresponding volume change.

$$\ln \frac{T_2}{T_1} = \left(\frac{\Delta V}{\Delta H}\right)(P_2 - P_1)$$

For phase transitions between a condensed phase and a gas, ΔV is V gas $= \frac{RT}{P}$ which gives the following **CLAUSIUS–CLAPEYRON EQUATION:**

$$\frac{dp}{dT} = \Delta H \left(\frac{P}{RT^2}\right)$$

$$\ln \frac{P_2}{P_1} = \left(\frac{-\Delta H}{R}\right)\left(\frac{1}{T_2} - \frac{1}{T_1}\right)$$

CHAPTER 6

THE SECOND AND THE THIRD LAWS OF THERMODYNAMICS

6.1 HEAT ENGINES

An engine converts energy from one form into another. A heat engine converts heat to work. The following figure shows the operation of a general heat engine.

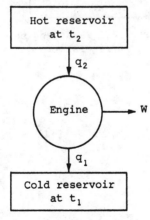

Fig. 6.1 Heat engine.

$$-w = q_2 - q_1$$

$$e = \frac{w_{output}}{q_{input}}$$

where e is the efficiency of the engine, w_{output} is the work done by the engine (system), and q_{input} is heat supplied to the engine (system).

$$e = -\frac{w}{q_2} = 1 - \frac{q_1}{q_2}$$

A refrigerator is an engine used to pump heat from a cold to a hot reservoir. A refrigerator, or a heat pump is a heat engine functioning in reverse. The following figure shows how the refrigerator operates.

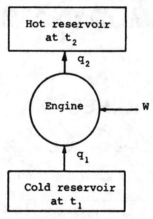

Fig. 6.2   Heat engine as refrigerator.

$$q_2 = -w + q_1$$

$$w = q_1 - q_2$$

$$e = \frac{w}{q_1} = 1 - \frac{q_2}{q_1}$$

6.2 CARNOT CYCLE

A carnot engine is an idealized heat engine that follows the cyclic process in Figure 6.3. The process consists of four steps performed on an ideal gas: (1) a reversible

34

isothermal expansion at T_h, (2) a reversible adiabatic expansion to T_c, (3) a reversible isothermal compression at T_c, and (4) a reversible adiabatic compression to T_h.

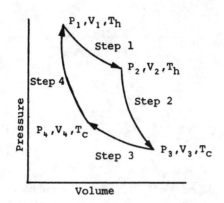

Fig. 6.3 Carnot Cycle

$$V_3 = V_2 \left(\frac{T_h}{T_c} \right)^{\frac{C_v}{nR}} \qquad \text{for adiabatic expansion}$$

$$q_1 = -w_1 = nRT_h \ln \frac{V_2}{V_1} = nRT_h \frac{\ln P_1}{P_2} \qquad \begin{array}{l}\text{for isothermal} \\ \text{expansion}\end{array}$$

$$e_{carnot} = \frac{T_h - T_c}{T_h} \qquad H \cdot P.$$

where e_{carnot} is the efficiency for a carnot engine.

$$e_{ref} = \frac{T_h - T_c}{T_c} \qquad REF$$

where e_{ref} is the efficiency for a carnot refrigerator.

$$c_o = \frac{T_c}{T_h - T_c}$$

where C_o is the coefficient of performance which is the least possible work required to withdraw a given amount of heat. The coefficient of performance is also equal to

$$c = \frac{q_c}{-w}$$

where q_c is equal to q_1.

6.3 ENTROPY

$\frac{q_{rev}}{T}$ is a state function, and is expressed by the new function entropy, S.

$$dS = \frac{dq_{rev}}{T}$$

$$\Delta S = S_{fin.} - S_{init.} = \int_{init}^{fin} \frac{dq_{rev}}{T}$$

S is also an extensive property because it depends upon the amount of material present. It is a state function, since it does not depend on the path of the process.

For a complete carnot cycle,

$$\Sigma \frac{q_{rev}}{T} = 0$$

$$\Delta S = \int_{cycle} \frac{dq_{rev}}{T} = 0$$

$$\Delta S = \int_{cycle} \frac{-dw_{rev}}{T} = 0$$

For a reversible process

$$\Delta S(\text{system}) = -\Delta S(\text{surroundings})$$

$$\Delta S(\text{universe}) = \Delta S(\text{system}) + \Delta S(\text{surroundings}) = 0$$

For an irreversible process

$$\Delta S(\text{universe}) > 0$$

CLAUSIUS'S INEQUALITY

The efficiency of any reversible carnot cycle is more than the efficiency of an irreversible carnot cycle that operates between the same two temperatures.

$$\oint \frac{dq}{T} = \int_{\text{cycle}} \frac{dq}{T} = 0 \qquad \text{(for reversible process)}$$

$$\oint \frac{dq}{T} < 0 \qquad \text{(for irreversible process)}$$

$dq = 0$ and $\Delta S \geq 0$ for an isolated system. $\Delta S = 0$ only when the system is at equilibrium with its surroundings.

The entropy of an isolated system always increases during an irreversible process. The second law observed that spontaneous processes occur in the direction of increasing entropy in an isolated system. The second law also states that heat cannot flow from a cold to a hot body spontaneously and heat cannot be extracted from a hot source and turned entirely into work.

Entropy tends to a maximum at constant energy and energy tends to a minimum at constant entropy.

The entropy of a chemical reaction is determined by the equation

$$S(V) = k \ln W(V)$$

where k is the Boltzmann constant and W is the probability that every particle is in a given volume V (container) or the number of detailed molecular states corresponding to the given gross state.

$$\Delta S = \Sigma S_{\text{product}} - \Sigma S_{\text{reactant}}$$

6.4 ENTROPY CALCULATIONS AND ABSOLUTE ENTROPIES

Isothermal expansion of an ideal gas

$$\Delta S = nR \ln \frac{V_2}{V_1} = nR \ln \frac{P_1}{P_2}$$

$\Delta S > 0$ for an expansion and $\Delta S < 0$ for a compression.

For a condensed system

$$\Delta S(system) = \int_{P_1}^{P_2} -\left(\frac{\partial V}{\partial T}\right)_P dp$$

Heating at constant pressure

$$\Delta S = \int_{T_1}^{T_2} Cp \frac{dT}{T}$$

If the heat capacity is not a function of temperature)δT small

$$\Delta S = Cp \int_{T_1}^{T_2} \frac{dT}{T} = Cp \ln \frac{T_2}{T_1}$$

Heating at constant volume

$$\Delta S = \int_{T_1}^{T_2} \frac{(dq_{rev})v}{T} = \int_{T_1}^{T_2} \frac{dE}{T} = \int_{T_1}^{T_2} Cv \frac{dT}{T}$$

If the heat capacity is not a function of temperature)δT small

$$\Delta S = Cv \ln \frac{T_2}{T_1}$$

ΔS for phase transitions

$$\Delta S = \frac{\Delta H_t}{T_t}$$

where ΔH_t is the enthalpy of phase-transition, and T_t is the transition temperature.

In order to determine the entropy $S(T)$ at a temprature higher than the boiling point, the following equation is applied:

$$S(T) = S(o) + \int_0^{T_f} \left(\frac{C_p^{solid}}{T} \right) dT + \frac{\Delta H_{melt}}{T_f}$$

$$+ \int_{T_f}^{T_b} \left(\frac{C_p^{liq}}{T} \right) dT + \frac{\Delta H_{evap}}{T_b} + \int_{T_b}^{T} \left(\frac{C_p^{gas}}{T} \right) dT$$

ΔS for adiabatic processes

$$\Delta S(system) = 0$$

ΔS for isothermal mixing

$$\Delta S(system) = -R \sum_i n_i \ln x_i$$

where n_i is the number of moles, and x_i is the mole fraction of component i in the mixture respectively.

Entropy changes in the surroundings

$$dq_{rev}^{surr} \quad \text{and} \quad dq_{irrev}^{surr} \quad \text{are equal to } dE^{surr}$$

$$ds^{surr} = \frac{dq^{surr}}{T^{surr}}$$

$$\Delta S^{surr} = \frac{q^{surr}}{T^{surr}} = -nR\ln\left(\frac{V_f}{V_i} \right) = -\Delta S^{system}$$

where q^{surr} is the total amount of heat injected from the system to the surroundings.

For any reversible process with expansion work only:

$$dE = dq_{rev} - PdV$$

$$dE = TdS - PdV$$

$$\left(\frac{\partial E}{\partial S}\right)_V = T \qquad \left(\frac{\partial E}{\partial V}\right)_S = -P$$

$$dS = \left(\frac{\partial S}{\partial T}\right)_V dT + \left(\frac{\partial S}{\partial V}\right)_T dV$$

$$\left(\frac{\partial S}{\partial T}\right)_V = \frac{C_v}{T} \qquad \left(\frac{\partial S}{\partial V}\right)_T = \frac{1}{T}\left[P + \left(\frac{\partial E}{\partial V}\right)_T\right]$$

$$\left(\frac{\partial E}{\partial V}\right)_T = \frac{nRT}{V} - P = 0$$

$$dH = TdS + Vdp$$

$$\left(\frac{\partial S}{\partial T}\right)_P = \frac{Cp}{T}, \qquad \left(\frac{\partial H}{\partial P}\right)_T = V - T\left(\frac{\partial V}{\partial T}\right)_P$$

$$\left(\frac{\partial H}{\partial P}\right)_T = 0 \quad \text{for an ideal gas.}$$

Absolute entropies at any temperature are determined by calculating ΔS from the fiducial point.

$S^{\circ}_{298.15}$ is the absolute entropy at the standard conditions of 1 atm and 298.15°K (STP).

6.5 THE THIRD LAW OF THERMODYNAMICS

The absolute zero temperature is unattainable by any process in a finite number of idealized steps.

For an adiabatic and reversible process, $q_{rev} = 0$ and $\Delta S = 0$.

Consider the process $a(T') \rightarrow b(T'')$

$$S_b^o - S_a^o = \int_0^{T'} C_a \, d(\ln T) - \int_0^{T''} C_b \, d(\ln T)$$

where S_b^o and S_a^o are the entropies of b and a at $0\,^\circ K$; C_b and C_a are the respective heat capacities of products (b) and reactants (a).

The third law states that all perfect crystals have the same entropy at absolute zero.

The third law fiducial point applies to perfectly ordered pure crystals.

$$S^o = -R \sum_i x_i \ln x_i$$

$S_o^o > 0$ for non-perfect crystals. The standard entropies of gases are all almost the same and are generally bigger than the entropies of liquids and solids of comparable complexity.

The value of S_o^o for crystals whose molecules orient themselves in Ω ways is determined by the equation

$$S_o^o = nR \ln \Omega$$

The value of S_o^o determined by the above equation is large because the real crystal's molecules do not orient in a completely random manner due to their weak intermolecular forces and size.

The value of S_o^o for a glassy material lies between the value of S_o^o for the solid state and the value of S_o^o for the liquid state.

$$S_o^o = \Delta S^o (\text{mixing}) \quad \text{for ideal solid solutions}$$

41

$$S^\circ_T = S^\circ_0 + \sum_i^{\text{phases}} \int \frac{Cp^\circ}{T}\, dT + \sum_j^{\text{transitions}} \frac{\Delta H^\circ_j}{T_j}$$

$Cp = aT^3$ at low temperatures

ΔS°_T for a chemical reaction is determined by the equation

$$\Delta S^\circ_T(\text{reaction}) = \sum_i n_i S^\circ_{T,i} - \sum_j n_j S^\circ_{T,j}$$

products   reactants

where n_i and n_j are the stoichiometric coefficients of the balanced equation.

CHAPTER 7

WORK, FREE ENERGY AND CHEMICAL EQUILIBRIUM

7.1 MAXIMUM WORK

Regardless of whether the process is carried out reversibly or irreversibly, the value of dE does not change.

$$dq_{rev} = dE - dw_{rev} = TdS$$

$$dq_{irr} = dE - dw_{irr} < TdS$$

$$dw_{rev} - dw_{irr} < 0$$

MAXIMUM USEFUL WORK

In a carnot engine, the heat is converted into useful work.

$$w = w_{useful} + w_{exp}$$

where w_{exp} is the expansion work done against the surroundings, and w_{useful} is the net useful work. $w_{exp} = 0$ when the process is carried out at constant volume.

MAXIMUM USEFUL WORK AT CONSTANT S AND V: ENERGY

$$dE = dq + dw_{useful} + dw_{exp} = dq + dw_{useful} - PdV$$

For reversible process

$$dE = dw_{useful} + TdS - PdV$$

at constant S and V, dS and dV are zero;

$$dw_{useful} = dE \qquad \text{(const S and V)}$$

The energy of the system decreases when the system does work on the surroundings at constant T and V.

Maximum useful work at constant S and P; Enthalpy

$$dH = dq + dw_{useful} + dw_{exp} + PdV + Vdp$$

$$dH = dw_{useful} + TdS + Vdp \qquad \text{for reversible processes}$$

$$dH = dw_{useful} \qquad \text{at constant S and P}$$

For reversible and constant entropy processes

$$dE = dw_{useful} - PdV$$

$$dw_{useful} = (dE + PdV) = d(E + PV) = dH \qquad \text{at constant pressure}$$

Maximum useful work at constant T and V; Helmholtz energy

For reversible and constant volume processes,

$$dE = TdS + dw_{useful}$$

also, at constant temperature:

$$dw_{useful} = (dE - TdS) = d(E - TS) = dA$$

$$A = E - TS$$

The work function E - TS is called the Helmholtz free energy or the Helmholtz energy, A.

For an isothermal reversible process,

$$dA = dE - TdS = dq_{rev} - TdS + dw_{rev}$$

$$dA = dw_{rev} \qquad \text{at constant temperature}$$

At constant T and V, the maximum useful work obtained from a process is measured by the decrease in A. $(dA)_{T,V} = 0$ is an equilibrium state.

Maximum useful work at constant T and P; Free Energy AT constant T and P

$$dw_{useful} = d(E - TS + PV) = dG$$

The Gibbs function G is called the Gibbs potential, the Gibbs free energy, free energy or the Gibbs energy.

$$G = E - TS + PV$$

$$G = H - TS = A + PV$$

$$\boxed{dG = dw_{e, max}}$$ at constant P and T

$W_{e, max}$ is the maximum work from the system excluding expansion work and is called the net work. At constant T and P, the maximum useful work obtained from a process is measured by the decrease in free energy. The free energy G and the Helmholtz energy A are state functions.

At constant pressure and temperature, the process is spontaneous for a negative ΔG, non-spontaneous for a positive ΔG and at equilibrium for ΔG equal to zero.

At constant volume and temperature, the process is spontaneous for a negative ΔA, non-spontaneous for a positive ΔA and at equilibrium for ΔA equal to zero.

Free energy is an extensive property of the system. It varies with the amount of substance in a linear manner at constant T and P.

$$\Delta G = G_{products} - G_{reactants}$$

The reaction proceeds spontaneously from reactants to products when ΔG is negative. The reverse reaction will proceed spontaneously when ΔG is positive.

45

7.2 FREE ENERGY

$$\Delta G = \Delta H - T\Delta S \qquad \text{at constant T}$$

When a phase transition occurs, $T\Delta S = \Delta H$ and $\Delta G = 0$.

Free energy calculations

Isothermal expansion P_1 to P_2 of an ideal gas

$$\Delta G = nRT \ln \frac{P_2}{P_1} = nRT \ln \frac{V_1}{V_2}$$

For an isothermal expansion of an ideal gas,

$$\Delta A = \Delta G$$

Normal boiling

$$\Delta G = \Delta H_{vap} - \frac{T\Delta H_{vap}}{T} = 0$$

where ΔH_{vap} is the heat of vaporization.

Vapor pressure

$$\Delta G = RT \ln P$$

where P is the vapor pressure.

Dependence of free energy on chemical composition; the chemical potential

$$G = G(P, T, n_1, n_2, \ldots, n_i)$$

$$dG = \left(\frac{\partial G}{\partial T}\right)_{P, n_1, n_2, \ldots, n_i} dT + \left(\frac{\partial G}{\partial P}\right)_{T, n_1, n_2, \ldots, n_i} dP$$

$$+ \sum_i \left(\frac{\partial G}{\partial n_i}\right)_{P, T, n_j} dn_i$$

where n_j are all n's except n_i.

The term $\left(\dfrac{\partial G}{n_i}\right)_{P,T,nj}$ is the chemical potential, μ_i and is called the partial molar Gibbs function.

$$dG = Vdp - SdT + \sum_i \mu_i dn_i$$

The above equation is called the master equation of chemical thermodynamics.

μ_i, the chemical potential, is dependent upon the composition of the system. All elements in their standard states have zero chemical potentials. At equilibrium the chemical potential is a minimum.

$$dG = -SdT + VdP \qquad \text{at constant composition.}$$

$$dG = \sum_i \mu_i dn_i \qquad \text{at constant P and T.}$$

$$dE = TdS - PdV + \sum_i \mu_i dn_i$$

$$dH = TdS + VdP + \sum_i \mu_i dn_i$$

$$dA = -SdT - PdV + \sum_i \mu_i dn_i$$

$$\mu_i = \left(\frac{\partial E}{\partial n_i}\right)_{S,V,n_j} = \left(\frac{\partial H}{\partial n_i}\right)_{S,P,n_j} = \left(\frac{\partial A}{\partial n_i}\right)_{T,V,n_j}$$

$$= \left(\frac{\partial G}{\partial n_i}\right)_{P,T,n_j}$$

At constant pressure and temperature

$$\mu_i^a = \mu_i^b$$

where a and b are two phases in equilibrium.

$$G = G° + nRT \ln P$$

where $G°$ is the standard Gibbs function at 1 atm.

$$\mu = \mu° + RT \ln P$$

$$\Delta G° \text{(reaction)} = \Delta G°_F \text{(products)} - \Delta G°_F \text{(reactants)}$$

ΔG_F^o is the standard free energy of formation. At standard temperature and pressure the value of the free energy for elements is zero.

$$\Delta G^o(\text{reaction}) = -nFE^o$$

ΔG^o(reaction) = Standard free energy change of reactions
F = Faraday constant = 96.4856kJ mol^{-1} V^{-1}
n = number of moles of electrons in the balanced reaction
E^o = cell potential

7.3 THERMODYNAMIC RELATIONS

$$dE = TdS - PdV$$
$$dH = TdS + Vdp$$
$$dA = -SdT - PdV$$
$$dG = -SdT + Vdp$$

At constant S and V, $dE = 0$

At constant S and P, $dH = 0$

At constant T and V, $dA = 0$

At constant T and P, $dG = 0$

$$\left(\frac{\partial S}{\partial V}\right)_E = \frac{P}{T} \qquad\qquad \left(\frac{\partial S}{\partial P}\right)_H = -\frac{V}{T}$$

$$\left(\frac{\partial V}{\partial T}\right)_A = \frac{-S}{T} \qquad\qquad \left(\frac{\partial P}{\partial T}\right)_G = \frac{S}{V}$$

$$dE = \left(\frac{\partial E}{\partial V}\right)_S dV + \left(\frac{\partial E}{\partial S}\right)_V dS$$

$$dH = \left(\frac{\partial H}{\partial P}\right)_S dp + \left(\frac{\partial H}{\partial S}\right)_P dS$$

$$dA = \left(\frac{\partial A}{\partial V}\right)_T dV + \left(\frac{\partial A}{\partial T}\right)_V dT$$

$$dG = \left(\frac{\partial G}{\partial P}\right)_T dp + \left(\frac{\partial G}{\partial T}\right)_P dT$$

$$\left(\frac{\partial E}{\partial V}\right)_S = -P \qquad\qquad \left(\frac{\partial E}{\partial S}\right)_V = T$$

$$\left(\frac{\partial H}{\partial P}\right)_S = V \qquad\qquad \left(\frac{\partial H}{\partial S}\right)_P = T$$

$$\left(\frac{\partial A}{\partial V}\right)_T = -P \qquad\qquad \left(\frac{\partial A}{\partial T}\right)_V = -S$$

$$\left(\frac{\partial G}{\partial P}\right)_T = V \qquad\qquad \left(\frac{\partial G}{\partial T}\right)_P = -S$$

The following relations are called the Maxwell's equations.

$$\left(\frac{\partial T}{\partial V}\right)_S = -\left(\frac{\partial P}{\partial S}\right)_V$$

$$\left(\frac{\partial T}{\partial P}\right)_S = \left(\frac{\partial V}{\partial S}\right)_P$$

$$\left(\frac{\partial P}{\partial T}\right)_V = \left(\frac{\partial S}{\partial V}\right)_T$$

$$\left(\frac{\partial V}{\partial T}\right)_P = -\left(\frac{\partial S}{\partial P}\right)_T$$

The thermodynamic equation of state is written as follows:

$$\left(\frac{\partial E}{\partial V}\right)_T = T\left(\frac{\partial P}{\partial T}\right)_V - P$$

7.4 THERMODYNAMIC FUNCTIONS AND THEIR MATHEMATICAL MANIPULATION

The functions E, H, A, and G are called potentials. The Legendre transformations are used to replace extensive properties by an associate intensive properties. This transformation is summarized in Table 7.1.

Table 7.1. Summary of the Legendre transformations of one variable to another.

Old		New
$v_1, v_2, \ldots, v_n$	Variable	$u_1, u_2, \ldots, u_n$
$F = F(v_1, v_2, \ldots, v_n)$	Function	$G = G(u_1, u_2, \ldots, u_n)$
$u_i = \dfrac{\partial F}{\partial v_i}$	New variable	$v_i = \dfrac{\partial G}{\partial u_i}$
$G = \Sigma\, u_i v_i - F$	New defined function	$F = \Sigma\, u_i v_i - G$
$G = G(u_1, u_2, \ldots, u_n)$	New function	$F = F(v_1, v_2, \ldots, v_n)$

REDUCTIONS OF DERIVATIVES

$$C_p = \left(\frac{\partial H}{\partial T}\right)_P = T\left(\frac{\partial S}{\partial T}\right)_P \qquad \alpha = \frac{(\partial v / \partial T)_P}{V}$$

$$C_v = \left(\frac{\partial E}{\partial T}\right)_V = T\left(\frac{\partial S}{\partial T}\right)_V \qquad \beta = \frac{-(\partial v / \partial p)_T}{V}$$

$$\left(\frac{\partial E}{\partial V}\right)_T = -P + T\,\frac{\alpha}{\beta}$$

where

$$\frac{\alpha}{\beta} = \left(\frac{\partial P}{\partial T}\right)_V$$

$$\left(\frac{\partial E}{\partial V}\right)_T = 0 \qquad \text{for ideal gases}$$

$$\left(\frac{\partial E}{\partial V}\right)_T \neq 0 \quad \text{for real gases, liquids, or solids}$$

$$\left(\frac{\partial H}{\partial P}\right)_T = V(1 - T\alpha)$$

Because $\alpha = T^{-1}$ for ideal gases, $\left(\frac{\partial H}{\partial P}\right)_T = 0$ for ideal gases.

$$C_p - C_v = \frac{\alpha^2 VT}{\beta}$$

JOULE-THOMSON COEFFICIENT

$$\mu_{jt} = \left(\frac{\partial T}{\partial P}\right)_H = \frac{V(T\alpha - 1)}{C_p}$$

$\mu = 0$ for an ideal gas.

MATHEMATICAL MANIPULATIONS

The dependence of free energy on pressure.

$$\left(\frac{\partial G}{\partial P}\right)_T = V$$

$$\Delta G = \int dG = \int_{P_1}^{P_2} Vdp$$

$$\Delta G = (P_2 - P_1)V \quad \text{for solids or liquids}$$

$$\Delta G = nRT \ln\left(\frac{P_2}{P_1}\right) \quad \text{for ideal gases.}$$

The dependence of free energy on temperature

$$\left(\frac{\partial G}{\partial T}\right)_P = \frac{G - H}{T} = -S$$

$$\frac{\partial(G/T)}{\partial T}\Big|_P = \frac{-H}{T^2}$$

$$\left(\frac{\partial(G/T)}{\partial(1/T)}\right)_P = H$$

The above equations are called the Gibbs-Helmholtz equations.

$$\frac{1}{T^2} \Delta G_{T_2} = \frac{1}{T_1} \Delta G_{T_1} - \int_{T_1}^{T_2} \frac{\Delta H}{T^2}\, dT$$

$$\Delta G_T = J + kT - \Delta a T \ln T - \frac{\Delta b}{2} T^2 - \frac{\Delta c}{6} T^3 - \frac{\Delta d}{12} T^4$$

$$\Delta G_T = J' + k'T - \Delta a \ln T - \frac{\Delta b}{2} T^2 - \frac{\Delta c}{2} T^{-1}$$

$$\frac{\Delta (G_T^o - H_o^o)}{T} = \overset{\text{products}}{\underset{i}{\Sigma}}\, n_i \left\{ \frac{G_T^o - H_o^o}{T} \right\}_i - \overset{\text{reactants}}{\underset{j}{\Sigma}}\, n_j \left\{ \frac{G_T^o - H_o^o}{T} \right\}_j$$

$$\Delta G_T^o(\text{reaction}) = T \left[\frac{\Delta (G_T^o - H_o^o)}{T} \right] + \Delta H_o^o(\text{reaction})$$

The value of ΔH can be determined from the slope of the curve obtained by plotting $\frac{\Delta G}{T}$ as a function of T^{-1}.

ENTROPY

$$\left(\frac{\partial G}{\partial P} \right)_T = V \qquad \left(\frac{\partial G}{\partial T} \right)_P = -S$$

$$\left(\frac{\partial S}{\partial P} \right)_T = -\left(\frac{\partial V}{\partial T} \right)_P = -\alpha V$$

where α is the coefficient of thermal expansion.

$$dS = \frac{C_v\, dT}{T} \qquad \text{at constant volume}$$

$$\Delta S = \int_{T_1}^{T_2} C_v\, d(\ln T) \qquad \text{at constant volume}$$

$$\Delta S = \int_{T_1}^{T_2} C_p\, d(\ln T) \qquad \text{at constant pressure}$$

CHAPTER 8

THE EQUILIBRIUM CONSTANTS FOR GAS REACTIONS

8.1 EQUILIBRIUM CONSTANTS FOR IDEAL GAS REACTIONS

$$aA + bB \rightleftharpoons cC + dD$$

$$\boxed{\Delta G = \Delta G^\circ + RT \ln k}$$

$$k = \frac{[C]^c [D]^d}{[A]^a [B]^b}$$

k is the thermodynamic equilibrium constant at equilibrium. $\Delta G = 0$ at equilibrium.

$$\ln k = -\frac{\Delta G^\circ}{RT}$$

$$k = e^{-\Delta G^\circ / RT}$$

where ΔG° is the standard state free energy.

$$\frac{d(\ln k)}{d(1/T)} = \frac{-\Delta H^\circ (T)}{R}$$

$$\ln\left(\frac{k_{T_2}}{k_{T_1}}\right) = \frac{1}{R} \int_{T_1}^{T_2} \frac{\Delta H^\circ}{T^2}\, dT \qquad \text{when} \quad \Delta H^\circ \text{ is a function of } T$$

$$\ln\left(\frac{k_{T_2}}{k_{T_1}}\right) = \frac{\Delta H^\circ}{R}\left(\frac{1}{T_1} - \frac{1}{T_2}\right)$$

when ΔH° is constant and is independent of temperature.

$-\dfrac{\Delta H^\circ}{R}$ is the slope of the curve obtained by plotting $\ln k$ versus $\dfrac{1}{T}$.

$$-\Delta G^\circ = RT \ln k_p$$

$$k_p = \left[\frac{(P_C^{\,eq})^c (P_D^{\,eq})^d}{(P_A^{\,eq})^a (P_B^{\,eq})^b}\right]$$

where k_p is the equilibrium constant and P_A^{eq} is the partial pressure of A at equilibrium and so on.

$$\left(\frac{\partial k_p}{\partial P}\right)_T = 0$$

$$k_p = \left\{\frac{(X_C)^c (X_D)^d}{(X_A)^a (X_B)^b}\right\}_{eq} (P)^{\Delta n}$$

$$k_p = k_x\, p^{\Delta n}$$

where k_x is the equilibrium constant in terms of mole fractions, p is the total pressure of the gases including inert gases if present and

$$\Delta n = \sum_i n_i(\text{products}) - \sum_j n_j(\text{reactants})$$

$$\left(\frac{\partial \ln k_x}{\partial \ln p}\right)_T = -\Delta n$$

The equilibrium constant k_x increases as the pressure is raised, for a negative Δn, k_x decreases as the pressure is raised, for a positive Δn. The value of k_x is not affected by the pressure when $\Delta n = 0$.

$$k_p = (RT)^{\Delta n} k_c$$

where k_c is the equilibrium constant in terms of molarities.

In electrochemical reactions, the equation below is used to calculate the equilibrium constant:

$$\ln k = \frac{nF \varepsilon^{\circ}}{RT}$$

8.2 THERMODYNAMIC PROPERTIES OF A VAN DER WAALS GAS

Expansion work

$$\left(P + \frac{n^2 a}{V^2}\right)(V - nb) = nRT$$

$$P = \frac{nRT}{V - nb} - \frac{n^2 a}{V^2}$$

For a reversible expansion work

$$-w = \int_{V_1}^{V_2} P_{ex} dV = \int_{V_1}^{V_2} \left(\frac{nRT}{V - nb} - \frac{n^2 a}{V^2}\right) dV$$

For a reversible expansion and isothermal process

$$-w_{vdw} = nRT \ln\left(\frac{V_2 - nb}{V_1 - nb}\right) + n^2a\left(\frac{1}{V_2} - \frac{1}{V_1}\right)$$

where V_1 and V_2 are the initial and final volumes.

ΔE for a van der Waals gas, (vdw) gas

$$\left(\frac{\partial E}{\partial V}\right)_T = T\left(\frac{\partial P}{\partial T}\right)_V - P$$

$$\left(\frac{\partial E}{\partial V}\right)_T = \frac{nRT}{(V - nb)} - \left[\frac{nRT}{V - nb} - \frac{n^2a}{V^2}\right]$$

$$\left(\frac{\partial E}{\partial V}\right)_T = \frac{n^2a}{V^2}$$

For an isothermal expansion process of a vdw gas, the energy change is

$$\Delta E = \int_{V_1}^{V_2} \left(\frac{n^2a}{V^2}\right) dV = -n^2a\left(\frac{1}{V_2} - \frac{1}{V_1}\right)$$

As a van der Waals gas is expanded isothermally, the value of its internal energy increases. This increase is proportional to 'a'. The variable 'a' is related to the attractive forces between molecules.

$$\Delta E = q + w$$

For a reversible isothermal expansion of a vdw gas the heat q is

$$q = \Delta E - w = nRT \ln\left(\frac{V_2 - nb}{V_1 - nb}\right)$$

For an adiabatic process, $q = 0$ and $dE = dw$.

$$dE = \left(\frac{\partial E}{\partial V}\right)_T dV + \left(\frac{\partial E}{\partial T}\right)_V dT$$

$$= \left(\frac{\partial E}{\partial V}\right)_T dV + C_v dT = dw = -PdV$$

$$dE - dw = 0$$

$$\left(\frac{nRT}{V - nb} - \frac{n^2a}{V^2} \right) dV + C_v dT = 0$$

ΔG for a van der Waals gas

$$dG = Vdp - SdT$$

$$dp = \left[\frac{-nRT}{(V - nb)^2} + \frac{2n^2a}{V^3} \right] dV$$

For a reversible isothermal expansion of a vdw gas, $dT = 0$.

$$dG = Vdp = \left[\frac{-nVRT}{(V - nb)^2} + \frac{2n^2a}{V^2} \right] dV$$

8.3 FUGACITY

Fugacity is a kind of equivalent pressure and is defined by the equation

$$d\mu = d\bar{G} = RT \, d(\ln f)$$

$$d\mu_i = d\bar{G}_i = RT \, d(\ln f_i)$$

f_i is the fugacity of the ith component.

The standard state for ideal gases is chosen at unit pressure. The standard state for real gases is chosen at unit fugacity. The standard value for the chemical potential of a real gas is at a fugacity of 1 atm. The standard state is a theoretical state in which the gas behaves ideally.

$$\mu_i - \mu_i^{\,\circ} = RT \ln \left(\frac{f_i}{f_i^*} \right)$$

f_i^* is comparable to $P_i^{\,\circ}$. The fugacity is an idealized pressure. The fugacity is equal to the pressure for ideal gases.

$$\frac{f}{P} \to 1 \quad \text{as} \quad P \to 0$$

Evaluating fugacity for real gases

$$\Delta \overline{G} = \int_{P_1}^{P_2} \overline{V} dp = RT \ln \frac{f_2}{f_1}$$

where $\overline{V}$ is the molar volume.

$$\left[\frac{\partial \ln(f/P)}{\partial P} \right]_T = \frac{-\alpha}{RT}$$

$$\alpha = \left(\frac{RT}{P} - \overline{V} \right)$$

α measures the non-ideality of the gas.

$$\ln f = \ln P - \frac{1}{RT} \int_0^P \alpha \, dp$$

The values of fugacities for real gases are equal to the areas under the curves obtained by plotting α versus pressure.

$$RT \ln \frac{f}{P} = \int_0^P \left(V - \frac{RT}{P} \right) dp$$

$$RT \ln \frac{f}{P} = -RT \ln [P(v - b)]_0^P + \left. \frac{RTb}{v - b} \right]_0^P - \left. \frac{2a}{\overline{V}} \right]_0^P$$

$$\ln f = \ln \frac{RT}{V - b} + \frac{b}{V - b} - \frac{2a}{RTV}$$

$$f = p \exp \int_0^P \frac{(z - 1)}{P} \, dp$$

where z is a function of temperature and pressure and is equal to $z = \frac{P\overline{V}}{RT}$.

$$f = P \exp \int_0^P (b/RT)dp = P \exp(Pb/RT)$$

When z is less than one, the fugacity is less than the pressure. When z is greater than one at higher pressures, the fugacity is greater than the pressure.

$$\gamma = \frac{f}{P} \qquad f = \gamma p$$

where γ is the fugacity coefficient.

$$\mu(P) = \mu^{\circ} + RT \ln(P) + RT \ln\gamma$$

The term $RT\ln\gamma$ is a measure of the deviations from ideality. The value of $RT\ln\gamma$ is equal to the area under the curve obtained by plotting $\overline{V} - (RT/P)$ as a function of P, where $\overline{V}$ is the molar volume. $\ln\gamma$ is the area under the curve obtained by plotting $z - 1$ as a function of $\ln P_r$. the value of γ for gaseous mixture is the same as that for the pure gas under the same conditions.

8.4 EQUILIBRIUM CONSTANTS FOR REAL GAS REACTIONS

$$aA(g) + bB(g) \rightleftharpoons cC(g) + dD(g)$$

$$f_i = x_i f_i^*$$

f_i^* is the fugacity of the pure gas at the same total pressure and temperature, and x_i is the mole fraction.

$$\gamma_i = \frac{f_i}{x_i P}$$

P is the total pressure.

$$\gamma_i = \frac{x_i f_i^*}{P} = \frac{f_i^*}{P} = \gamma_i^*$$

$$k_f = \frac{(f_C)^c (f_D)^d}{(f_A)^a (f_B)^b}$$

The equation for k_f is valid for ideal and non-ideal gases.

$$k_f = \left\{ \frac{(\gamma_C)^c (\gamma_D)^d}{(\gamma_A)^a (\gamma_B)^b} \right\} \left\{ \frac{(P_C)^c (P_D)^d}{(P_A)^a (P_B)^b} \right\}$$

$$k_f = k\gamma\, kp$$

$$\Delta \overline{G}^\circ = -RT \ln k$$

$$k = \left\{ \frac{(\gamma_C)^c (\gamma_D)^d}{(\gamma_A)^a (\gamma_B)^b} \right\}_e \left\{ \frac{(X_C)^c (X_D)^d}{(X_A)^a (X_B)^b} \right\}_e$$

$$k = k\gamma\, kx$$

where $k\gamma$ is the equilibrium constant expressed in terms of activity coefficients and kx in terms of mole fractions.

CHAPTER 9

PHASE EQUILIBRIA

9.1 STABILITY OF PHASES

The intensive properties of a homogeneous system are uniform throughout the system. A heterogeneous system is a system that is made up of two or more homogeneous parts with sudden changes in properties occuring at the boundaries of these parts.

A phase is homogeneous and has a uniform composition and physical state throughout the system. Different phases of matter refer to bodies with different compositions or states.

The minimum number of different chemical species needed to specify the chemical composition of the phase is called the number of components in a phase. A pure phase is the phase that contains only one component.

The number of degrees of freedom is the number of intensive variables of the system that can independently vary without changing the number of phases present.

CONDITIONS FOR THE STABILITY OF A PURE PHASE

The condition for equilibrium at constant volume and entropy is the minimum energy, $\delta E > 0$.

The temperature increases as the entropy of a stable phase increases. The pressure decreases as the volume increases.

For a two phase system consisting of the two subsystems 'a' and 'b', the total energy is equal to

$$E = E^a + E^b$$

If the temperature, pressure and chemical potentials are the same in both phases, then the system is in equilibrium. The condition for equilibrium is $dS = 0$ because the energy $E^a + E^b$ is constant.

$$dS = 0 = \left(\frac{1}{T^a} - \frac{1}{T^b} \right) dE^a + \left(\frac{P^a}{T^a} - \frac{P^b}{T^b} \right) dV^a$$

$$+ (\mu_i{}^a - \mu_i{}^b) dx_i{}^a$$

$T^a = T^b$, $P^a = P^b$ and $\mu_i{}^a = \mu_i{}^b$ at equilibrium.

9.2 THE PHASE RULE

The phase rule states that the total number of degrees of freedom equals the number of components less the number of phases plus 2.

$$\boxed{f = C - P + 2}$$

where f is the number of degrees of freedom, C is the number of components, and P is the number of phases.

$$v_{tot} = PC + 2$$

where v_{tot} is the total number of variables. Every written equation that relates the system variables reduces v_{tot} by 1.

9.3 THE ONE-COMPONENT SYSTEM

For a one-component system, $c = 1$ and $f = 3 - P$. When only one phase is present, $f = 2$, and P and T can vary independently. When two phases are present, $f = 1$, and only one of the variables will vary independently. When three phases are in equilibrium, $f = 0$ and none of the variables will vary independently.

$$f = 3 - P$$

$P = 1$;　　$f = 2$　　　　(bivariant system)

$P = 2$;　　$f = 1$　　　　(univariant system)

$P = 3$;　　$f = 0$　　　　(invariant system)

The point at which all three phases are in equilibrium is called the triple point. The point at which the liquid and solid are in equilibrium at 1 atm pressure is called the standard melting point (mp).

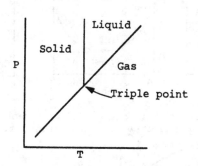

Fig. 9.1　One-component phase diagram.

9.4 THE TWO-COMPONENT SYSTEM

$F = 4 - P$ for a two-component system. When one phase is present, three variables are needed to describe a system. The temprature and composition are two of the independently varying variables holding the pressure constant.

$$f = 4 - P$$

P = 1;	f = 3	(trivariant system)
P = 2;	f = 2	(bivariant system)
P = 3;	f = 1	(univariant system)
P = 4;	f = 0	(invariant system)

LIQUID-LIQUID PHASE DIAGRAMS

In Figure 9.2, one or both of the components have solidified before reaching the point b, which is the lower consolute temperature, and one or both components have vaporized before reaching the point a, which is the upper consolute temperature.

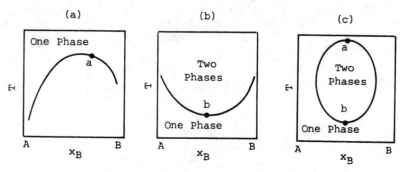

Fig. 9.2   There is a single one-phase area and a single two-phase area in these diagrams.

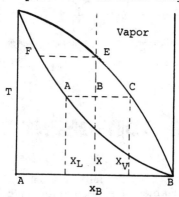

Fig. 9.3 Liquid-liquid phase diagram.

Figure 9.3 shows a typical phase diagram for liquid-liquid mixture.

The compositions of the liquid and vapor phases that are in equilibrium are given by the ends of the horizontal lines called tie lines. The points E and F are joined by a tie line.

$$x(n_L + n_V) = x_L n_L + x_V n_V$$

where n_L and n_V are the total number of moles of material in the liquid and vapor phases, respectively; x_L and x_V are the mole fractions of the liquid and vapor phases.

$$\frac{n_L}{n_V} = \frac{x_V - x}{x - x_L} = \frac{BC}{BA}$$

The lever rule states that the distances of the phase lines for the overall composition are inversely proportional to the relative amounts of phases. The lever rule is expressed by the above equation.

The diagram in Figure 9.3 is used to determine the number of theoretical equivalent plates, TEP, in a distillation column. In a TEP, which is a simple distillation step, equilibrium between the solution and vapor is established and the vapor is condensed to a liquid of different composition.

Liquid-Vapor Phase Diagrams

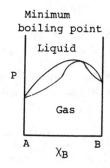
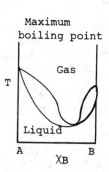

Fig. 9.4 Liquid-vapor P-χ diagram at constant T and T-χ diagram at constant P.

A maximum in the T-X diagram matches a minimum in the P-X diagram. When the liquid and the vapor compositions are the same in a solution, this solution is called an azeotropic solution. An azeotropic solution is inseparable into its components by distillation because its composition does not change upon boiling.

When a non-azeotropic solution is distilled in a minimum boiling point system, the vapor will direct towards the azeotropic concentration and the residual liquid towards one of the pure components. In a maximum boiling point system, the vapor will direct towards one of the pure components and the residual liquid towards the azeotropic concentration.

An isopleth is a line of constant overall composition.

Liquid-Solid Phase Diagrams; Simple Eutectics

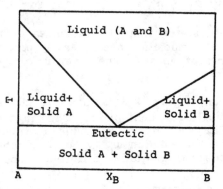

Fig. 9.5 A typical liquid-solid phase diagram.

Eutectic is the minimum point in the freezing-point curve. The liquid solution, the pure solid A and the pure solid B phases are in equilibrium at the eutectic point. A liquid mixture with the eutectic composition solidifies at the lowest temperature of the mixture. When a sample has the eutectic composition, cooling proceeds at constant rate until the eutectic temperature is achieved.

The liquid-solid phase diagrams are plotted using the cooling-curve measurements at different concentrations. For a pure compound or at the eutectic composition, the cooling curves consist of a plateau or "arrest." The cooling curve consists of a "break" or an "arrest" at other compositions. An "arrest" is the point at which the remaining liquid solidifies at the eutectic composition. A "break" occurs at the point where the solid begins to solidify and the liquid changes composition.

The solubility of a solid in a liquid is expressed in terms of the temperature variation by the equation

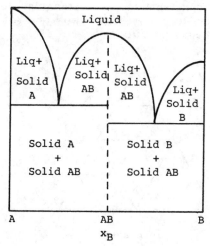

Fig. 9.6 Compound formation phase diagram

$$\ln X_B = \frac{\Delta H_{fus}}{R} \left(\frac{1}{T^o_m} - \frac{1}{T} \right)$$

where
X_B = The mole fraction of solute

ΔH_{fus} = The heat of fusion of the solute

T^o_m = The melting point of the solute

T = The temperature

COMPOUND FORMATION

Incongruent melting occurs when A_2B melts and equilibrates with a liquid mixture of A and B. The temperature of the following phase reaction (or peritectic reaction), that occurs at the melting point is the peritectic point.

$$A_2B(s) \rightleftharpoons A_{(L)} + A_{(L)} + B_{(L)}$$

9.5 THE THREE-COMPONENT SYSTEM

f = 5 - P for a three-component system. For a one phase system, four variables are used to specify the system. T and P are kept constant and a rectangular ternary-phase diagram is used to represent two out of the three compositions for the components.

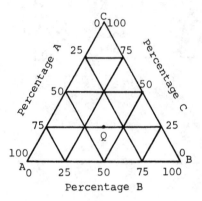

Fig. 9.7 Three component phase diagram.

The point Q represents a system with 40% A, 35% B, and 25% C. The three components in the above phase diagram are completely miscible in all proportions.

$$\chi_A + \chi_B + \chi_C = 1$$

In Figure 9.10 is shown a phase diagram for a ternary system with two phases: two conjugate ternary solutions, each containing the three components.

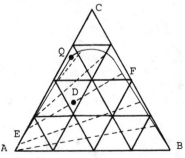

Fig. 9.8 Three-component two phases diagram.

The line EDF is a tie line. The lever rule is applicable to this system.

$$\frac{E}{F} = \frac{DF}{DE}$$

The point Q is the point at which the length of the tie line is very small, and it is called the critical or the plait point. There is a single phase system beyond the point Q.

(1)   (2)

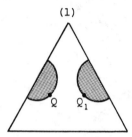 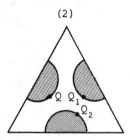

Fig.9.9   Three component phase diagrams for two pairs and three pairs of immiscible components.

In Figure 9.9 (1), the phase diagram is for two pairs of immiscible components. In Figure 9.9 (2), the phase diagram is for three pairs of immiscible components.

The mutual solubilities decrease and areas of the immiscible regions start to overlap as the temperature decreases. The merging of these immiscible regions is called encroachment.

9.6 FRACTIONAL DISTILLATION

Fractional distillation is used to separate liquids with different boiling points. Fractional distillation involves the repeating of the boiling and condensation cycle. A basic diagram of the distillation column is shown in Figure 9.10.

The returning liquid to the column is called the reflux and is directed by the condenser. The takeoff is the material removed from the column. The reflux ratio is the ratio of the material removed to the material returned to the pot. The amount of liquid forced through the still is the throughput. The amount of liquid needed to coat the inside of the column is the holdup. Rectification is a high degree of liquid-vapor contact.

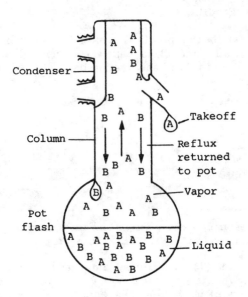

Fig. 9.10. A basic diagram of the distillation column.

In a bubble-cap column, each plate provides a surface for liquid-vapor equilibrium. Plates have lower temperatures and contain the liquid that is richer in more volatile components. Each plate accomplishes a degree of separation expressed by a separation step called a theoretical plate. High stills with large amounts of plates have high separation capacities.

Packing increases the liquid-vapor contact in the column. The packed distillation column is used to separate liquids with boiling points as close to each other as .5 - 1°C. The spinning-band column is a highly efficient distilling system.

The number of theoretical plates in a column is expressed in terms of HETP, or height equivalent to a theoretical plate.

$$HETP = \frac{\text{Length of the column}}{\text{number of theoretical plates}}$$

70

$$n = \frac{\log_{10}[(X_A/Y_A)(Y_B/X_B)]}{\log_{10}k} - 1$$

where n = the number of theoretical plates

X_A = % low-boiler in the head

Y_A = % low-boiler in the pot

Y_B = % high-boiler in the pot

X_B = % high-boiler in the head

k = the ratio of the vapor pressure of the low-boiler to the vapor pressure of the high-boiler.

9.7 SLOPES ON A PHASE DIAGRAM: THE CLAPEYRON EQUATION

$$\mu_1^a - \mu_2^a = \mu_1^b - \mu_2^b$$

$$d\overline{G}^a = d\overline{G}^b = (\overline{V}^a - \overline{V}^b)dp = (\overline{S}^a - \overline{S}^b)$$

The Clapeyron equation is

$$\frac{dP}{dT} = \frac{\Delta S}{\Delta V}$$

where ΔV is the molar volume change for phase transition, and ΔS is the molar entropy change for transition.

The solid-liquid boundary

$$\Delta S_{melt} = \frac{\Delta H_{melt}}{T_f}$$

where ΔS_{melt} is the molar entropy change on melting, ΔH_{melt} is the molar heat of fusion and T_f is the melting temperature.

$$\frac{dP}{dT} = \frac{\Delta H_{melt}}{T_f \Delta V_{melt}}$$

where ΔV_{melt} is the molar volume change on melting.

$$P_2 = P_1 + \left(\frac{\Delta H_{melt}}{\Delta V_{melt}}\right) \ln\left(\frac{T_2}{T_1}\right)$$

The above equation is the equation of the solid-liquid equilibrium curve. ΔH_{melt} and ΔV_{melt} are independent of T and P.

The liquid-gas boundary

$$\frac{dP}{dT} = \frac{\Delta H_{evap}}{T \Delta V_{evap}}$$

where ΔH_{evap} is the molar enthalpy of vaporization and ΔV_{evap} is the molar volume change on vaporization.

The following equation is the Clausius-Clapeyron equation:

$$\frac{d(\ln P)}{dT} = \frac{\Delta H_{evap}}{RT^2}$$

where ΔH_{evap} is independent of T and P.

$$\ln\left(\frac{P_2}{P_1}\right) = \frac{-\Delta H_{evap}}{R}\left(\frac{1}{T_2} - \frac{1}{T_1}\right)$$

The term $\frac{-\Delta H}{R}$ is equal to the slope of the curve obtained from plotting lnP as a function of 1/T.

The gas-solid boundary

$$\frac{d(\ln P)}{dT} = \frac{\Delta H_{sub}}{RT^2}$$

where ΔH_{sub} is the molar enthalpy of sublimation.

$$\ln\left(\frac{P_2}{P_1}\right) = \frac{-\Delta H_{sub}}{R}\left(\frac{1}{T_2} - \frac{1}{T_1}\right)$$

CHAPTER 10

COLLIGATIVE PROPERTIES OF IDEAL SOLUTIONS

10.1 SOLUTIONS

The chemical potential $\mu_i = \left(\dfrac{\partial G}{\partial n_i}\right)_{P,T}$ is adjusted by altering the boiling point or freezing point which is achieved by introducing another component to form a solution. A solution is a homogeneous mixture with a single phase of more than one component. Miscible substances are thoroughly soluble in each other.

$$X_A = \frac{n_A}{n_A + n_B + n_C + \ldots}$$

where X_A is the mole fraction of component A and n_A, n_B, and n_C are the number of moles of components A, B, and C, respectively.

Molarity, M, is one of the volume concentration units and it is defined as the number of moles of solute per liter of solution. Molality, m, is another concentration unit and is defined as the number of moles of solute per 1000g of solvent. One last concentration unit, normality, N, is defined as the number of equivalents of solute per liter of solution.

$$\text{Normality, } N = \left(\frac{\text{eq solute}}{\text{dm}^3 \text{ soln}}\right)$$

$$\boxed{\begin{array}{l} \text{Molarity, M} = \left(\dfrac{\text{mole solute}}{\text{liter soln}} \right) \\[4mm] \text{Molality, m} = \dfrac{\text{mole solute}}{\text{kg solvent}} \end{array}}$$

$$\chi_B = \frac{m_B}{(1000/M_A) + m_B} = \frac{m_B M_A}{1000 + m_B M_A}$$

where M_A is the molar mass of A and m_B is the molality of B.

The mass percent of the solvent, C_A, is

$$C_A = (100 - C_B)$$

where C_B is the mass percent of the solute.

$$w_A = .01 C_A \rho \qquad\qquad w_B = .01 C_B \rho$$

where w_A is the mass of the solvent in g/cm^3, w_B is the mass of the solute in g/cm^3 and ρ is the density at a concentration C_B.

The volume of solution containing 1000g of solvent per cm^3 is

$$V = \frac{1000}{w_A}$$

$$m_B = \left(\frac{1000}{w_A} \right) \left(\frac{w_B}{M_B} \right)$$

where M_B is the molecular weight of solute.

$$m_B = \frac{1000 C_B}{M_B (1 - C_B)}$$

where C_B is the mass percent of solute.

10.2 RAOULT'S LAW: THE IDEAL SOLUTION

Raoult's Law is expressed by the equation:

$$P_i = \chi_i P_i^o$$

where P_i is the partial pressure of component 'i' in the vapor phase above the solution, P_i^o is the vapor pressure of the pure component 'i' at that temperature, and χ_i is the mole fraction of the component in the solution.

Ideal solutions obey Raoult's law throughout the solution. Solutions with components that are chemically similar obey Raoult's law. The vapor pressure of the solute in an ideal solution agrees with Raoult's law.

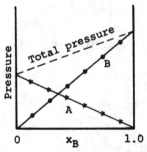

Fig. 10.1 The total pressure and partial pressures of components in ideal solution.

The following equations are applicable to both ideal and non-ideal solutions.

$$\mu(L) = \mu(g)$$

$$\mu(L) = RT \ln f$$

where f is the fugacity of the vapor in equilibrium with the liquid.

$$f = \chi f^o$$

75

where f° is the fugacity of the pure liquid.

10.3 PARTIAL MOLAR QUANTITIES

The increase in volume that results from adding 1 mole of a substance to a large amount of solution is called the partial molar volume of the substance in a mixture.

$$\Delta V = \overline{V}_A \Delta n_A$$

where ΔV = the increase in volume observed

Δn_A = the amount added

$\overline{V}_A$ = the partial molar volume of A.

$$\overline{V}_A = \left(\frac{\partial V}{\partial n_A} \right)_{T,P,n_B}$$

The partial molar quantities for the various extensive state functions are:

$$\overline{S}_A = \left(\frac{\partial S}{\partial n_A} \right)_{T,P,n_B}$$

$$\overline{H}_A = \left(\frac{\partial H}{\partial n_A} \right)_{T,P,n_B}$$

$$\overline{G}_A = \left(\frac{\partial G}{\partial n_A} \right)_{T,P,n_B}$$

$$\overline{V}_A = \left(\frac{d\overline{G}_A}{\partial P} \right)_T = \left(\frac{d\mu_A}{\partial P} \right)_T$$

$$-\overline{S}_A = \left(\frac{\partial \mu_A}{T} \right)_P$$

$$\mu_A = \overline{G}_A = \left(\frac{\partial G}{\partial n_A}\right)_{P,T,n_B}$$

A partial molar quantity is an intensive property of the system because it is independent of the size of the system.

$$V = n_1\overline{V}_1 + n_2\overline{V}_2 + n_3\overline{V}_3 + \dots$$

where n_i = the number of moles of component i

$\overline{V}_i$ = the partial molar volume of component i

V = the volume of solution.

The Gibbs-Duhem equation for a two-component solution is expressed by the equation

$$d\overline{V}_A = -\left(\frac{n_B}{n_A}\right) d\overline{V}_B$$

$$d\overline{V}_A = -\left(\frac{\chi_B}{\chi_B - 1}\right) d\overline{V}_B$$

$$n_A d\overline{V}_A = -n_B d\overline{V}_B$$

$$G = n_A \mu_A + n_B \mu_B + \dots$$

$$= \Sigma n_i \mu_i$$

$$\Sigma n_i \mu_i = 0$$

The above equation is an expression of the Gibbs-Duhem equation.

$$\overline{V}_B = \left(\frac{\partial V}{\partial n_B}\right)_{n_A,P,T}$$

= slope of the tangent to the curve obtained from plotting the volume as a function of the solute concentration for an aqueous solution.

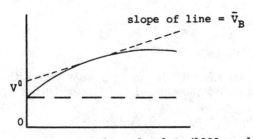

slope of line = $\bar{v}_B$

V^0

0

n_B = no. of moles of solute/1000g solvent

Fig. 10.2   Volume as a function of solute
concentration for an aqueous solution.

10.4 MIXING OF IDEAL SOLUTIONS

Volume change on mixing

$$\bar{V}_i = \bar{V}_i^o$$

where $\bar{V}_i^o$ is the molar volume of the pure component and $\bar{V}_i$
is the molar volume of the solution.

$$\Delta V = V_{fin} - V_{init} = 0$$

where ΔV is the volume change on mixing

$$\Delta V_{mixing} = 0$$

When mixing for an ideal solution, the volume does not
change.

Enthalpy change on mixing

Heat does not evolve on mixing two pure components to
form an ideal solution.

$$\Delta H_{mixing} = 0$$

Entropy and Free energy change on mixing

78

$$\Delta G(\text{mixing}) = G(\text{solution}) - G(\text{pure})$$

$$\Delta G(\text{mixing}) = RT \sum_i^{\text{components}} n_i \ln \chi_i$$

$$\Delta G(\text{mixing}) < 0$$

$$\Delta S(\text{mixing}) = \frac{\Delta H(\text{mixing}) - \Delta G(\text{mixing})}{T}$$

$$\Delta S(\text{mixing}) = \frac{-\Delta G(\text{mixing})}{T} = -R \sum_i^{\text{components}} n_i \ln \chi_i$$

$$\Delta S(\text{mixing}) > 0$$

10.5 DILUTE SOLUTIONS AND HENRY'S LAW

The vapor pressure of the solute in dilute solutions is proportional to the mole fraction. This relationship is expressed by the following Henry's law equation.

$$P_B = k \chi_B$$

where χ_B is the mole fraction of the solute and k_B is Henry's law constant.

For a dilute solution that obeys Henry's law

$$f_B = k \chi_B$$

Henry's law is also expressed by the equation

$$f_B = k'_m$$

where m is the molality, or moles of solute per 1000g of solvent.

An application of Henry's law is the distribution of a solute between two phases in equilibrium.

The chemical potential and the fugacity are the same for both phases

$$f_B^{\,1} = f_B^{\,2}$$

Henry's law of dilute solutions is

$$f_B^{\,1} = k_1 \chi_B^{\,1} \qquad \text{in phase 1}$$

$$f_B^{\,2} = k_2 \chi_B^{\,2} \qquad \text{in phase 2}$$

The ratio of the concentrations of the solute distributed between the two phases is determined by the Nernst distribution law equation.

$$\frac{\chi_B^{\,2}}{\chi_B^{\,1}} = \frac{k_1}{k_2} = k$$

where k is the distribution constant.

10.6 ACTIVITIES

$$\Delta \overline{G}_i = \overline{G}_i - G_i^{\circ} = \mu_i - \mu_i^{\circ} = RT \, \ln\left(\frac{f_i}{f_i^{\circ}}\right)$$

$$a = \frac{f}{f^{\circ}}$$

where $\frac{f}{f^{\circ}}$ is the relative fugacity and a is the activity.

$$\Delta \overline{G}_i = \mu_i - \mu_i^{\circ} = RT \, \ln a_i$$

where μ_i° is chemical potential in the standard state.

Activities of solvents

$$a_A = \frac{f_A}{f_A^{\circ}} = \frac{f_A^{\circ}}{f_A^{\circ}} = 1 \qquad \text{for a pure solvent.}$$

For an ideal gas

$$a_A = \frac{P_A}{P_A^o} \qquad a_A = \gamma_A \chi_A \quad \text{or} \quad \gamma_A = \frac{a_A}{\chi_A}$$

where γ_A is the activity coefficient.

$\gamma_A = 1$ and $\chi_A = 1$ for a pure solvent.

$$\gamma_A = \frac{f_A}{\chi_A f_A^o} \qquad \text{or} \qquad \gamma_A = \frac{P_A}{\chi_A P_A^o}$$

$$\mu_A - \mu_A^o = RT \ln(\gamma_A \chi_A)$$

Activities of solutes

For dilute solutions, the solute obeys Henry's law, which is

$$P_B = k_B \chi_B$$

where P_B is the vapor pressure of the solute, k_B is Henry's law constant, and χ_B is the mole fraction of the solute.

$$f_B = k_B \chi_B$$

$$a_B = \frac{f_B}{f_B^o} = \frac{P_B}{P_B^o}$$

At the standard hypothetical state for an ideal dilute solution

$$a_B = \chi_B \qquad \text{and} \qquad \gamma_B = 1$$

$$\gamma_B = \frac{a_B}{\chi_B} \qquad \gamma_B = \frac{P_B}{k_B \chi_B} \quad \text{or} \quad \gamma_B = \frac{f_B}{k_B \chi_B}$$

$$f_B = k_B' m_B \qquad P_B = k_B' m_B$$

$a_B = m_B$ and $\gamma_B = 1$ when Henry's law is valid.

$$\gamma_B = \frac{a_B}{m_B} = \frac{P_B}{k'_B m_B}$$ when Henry's law is not valid.

$$\mu_B - \mu_B^o = RT \ln a_B = RT \ln(\gamma_B m_B)$$

For gases $a = f$ and for ideal gases $a = P$. For pure solids and liquids $a = 1$ at the standard state. In dilute solutions for which Henry's law is valid, $a_B = m_B$. When Henry's law is not valid, $a_B = \gamma_B m_B$.

Equilibrium constant in terms of activities

For the reaction

$$aA(a_A) + bB(a_B) \rightleftharpoons cC(a_C) + dD(a_D)$$

the equilibrium constant at constant T and P is

$$k_a = \frac{(a_C)^c (a_D)^d}{(a_A)^a (a_B)^b}$$

where a_A, a_B, a_C and a_D are the activities for each species at equilibrium.

$$k_a = \frac{(\gamma_C m_C)^c (\gamma_D m_D)^d}{(\gamma_A m_A)^a (\gamma_B m_B)^b} = \frac{(\gamma_C)^c (\gamma_D)^d}{(\gamma_A)^a (\gamma_B)^b} \frac{(m_C)^c (m_D)^d}{(m_A)^a (m_B)^b}$$

$$k_a = k_\gamma k_m$$

The activity coefficient of the solutes are unity and $k_\gamma = 1$ when Henry's law is valid in a very dilute solution. Therefore, $k_a = k_m$.

An application of the Gibbs-Duhem equation to activities

The following equation is a form of the Gibbs-Duhem equation:

$$n_A d\overline{X}_A = -n_B d\overline{X}_B$$

$$X_A d(\ln a_A) + X_B d(\ln a_B) = 0$$

$$\ln a_A = - \int_{X_B=0}^{X_B} \frac{X_B}{1 - X_B} \, d(\ln a_B)$$

where a_A = the activity of the solvent

           X_A = the mole fraction of the solvent

           a_B = the activity of the solute

           X_B = the mole fraction of the solute.

$$\ln \frac{a_B}{X_B} = - \int_{X_B=0}^{X_B} \frac{X_A}{X_B} \, d \left[\ln \left(\frac{a_A}{X_A} \right) \right]$$

$$\ln \gamma_B = - \int_{X_B=0}^{X_B} \frac{X_A}{X_B} \, d(\ln \gamma_A)$$

$$X_A \, d \ln \gamma_A = - X_B \, d \ln \gamma_B$$

10.7 OSMOTIC PRESSURE

The external pressure necessary to stop the spontaneous flow of a pure solvent into a solution through a semipermeable membrane is the osmotic pressure, π.

$$P - P^o = \frac{RT X_B}{\overline{V}_A^o}$$

$P - P^o$ = osmotic pressure = π

    X_B = the mole fraction of the solute

    $\overline{V}_A^o$ = the partial molar volume of the solvent.

$$\pi = \frac{RT}{\overline{V}_A^o} \frac{n_B}{n_A} = n_B \frac{RT}{V}$$

where V – the total volume of the solution.

$$\pi = cRT$$

where c = the molar concentration of the solution.

$$\pi = \frac{gRT}{MW(V)}$$

where   g = the grams of solute

       MW = molecular weight of solute

Because osmosis is dependent on the amount and not the nature of the solute, it is a colligative property.

10.8 FREEZING-POINT DEPRESSION AND BOILING-POINT ELEVATION

FREEZING-POINT DEPRESSION

A solution freezes at a lower temperature than that of the pure solvent. The difference between the temperatures is called the freezing-point depression, ΔT_f.

$$\Delta T_f = \frac{RT_f^2}{\Delta H_{fus}} X_B$$

where $\Delta T = T_f - T$ = the freezing-point depression

       T_f = the freezing point of the pure solute

     ΔH_{fus} = the heat of fusion of the solute

       X_B = the mole fraction of the solute

$$\Delta T_f = k_f m_B$$

where k_f = the freezing-point depression constant

or the cryoscopic constant

m_B = the molal concentration of the solute.

$$k_f = \frac{RT_f^2 M_A}{1000 \, \Delta H_{fus}}$$

where M_A = the molar mass of the solvent.

The following equation indicates the relationship between the solubility of the solute and the temperature.

$$\ln \chi_B = \frac{-\Delta H_{fus}}{R} \left(\frac{1}{T} - \frac{1}{T_f} \right)$$

BOILING-POINT ELEVATION

Solutions containing non-volatile solutes have higher boiling points than those of pure solvents. The difference between the temperatures is called the boiling point elevation, ΔT_b.

$$\Delta T_b = k_b m_B$$

where ΔT_b = the boiling-point elevation

k_b = the boiling-point elevation constant or the ebullioscopic constant

m_B = the molal concentration of the solute.

$$k_b = \frac{RT_b^2 M_A}{1000 \, \Delta H_{vap}}$$

where T_b = the boiling point temperature

ΔH_{vap} = the heat of vaporization of the solvent

M_A = the molar mass of the solvent.

CHAPTER 11

IONS IN SOLUTIONS

11.1 CONDUCTIVITY

The motion of ions in solution is determined by their conductivity, which is their ability to conduct electricity.

$$\kappa = \frac{1}{\rho}$$

where κ is the conductivity or specific conductance on ohm^{-1} and ρ is the resistivity or specific resistance in ohms.

$$R = \frac{\rho L}{A} = \frac{L}{\kappa A}$$

where

R = the resistance of a material

L = the length or the diatance between electrodes

A = the surface area of an electrode.

The resistance to the flow of electricity of a cell is measured by using a bridge circuit similar to the one shown in Figure 11.1.

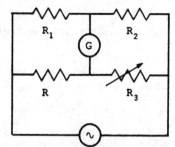

Fig. 11.1 A bridge circuit
to measure the resistance.

$$R = R_3 \frac{R_1}{R_2}$$

Ohm's law is expressed by the equation

$$I = \frac{V}{R}$$

where I = current

         V = potential drop across the circuit, voltage

         R = the resistance of the circuit.

The conductivity is dependent upon the total charge and
its size.

$$\Lambda = \frac{\kappa}{c}$$

where Λ = the molar conductance

         c = the concentration in equivalents per
            cubic centimeter.

$$\Lambda = \frac{\kappa}{1000c}$$

where Λ = the equivalent conductance

         c = the concentration in equivalents per dm^3.

$$\Lambda = \Lambda_0 - k\, c^{\frac{1}{2}}$$

where Λ_0 = the molar conductivity at infinite dilution

         k = a constant that is dependent upon the
            nature of the salt.

When Λ is plotted as a function of $c^{\frac{1}{2}}$, a linear plot is obtained for a strong electrolyte and a highly curved plot for a weak electrolyte.

The Kohlrausch's law of independent mobilities of ions in infinitely dilute solutions is expressed by the equation

$$\Lambda_0 = \lambda_+^o + \lambda_-^o$$

where λ_+^o = the molar conductivity of the cation

λ_-^o = the molar conductivity of the anion

Λ_0 = the molar conductivity at infinite dilution.

Weak electrolytes

$$n_+ = \alpha n \qquad n_- = \alpha n$$

where n_+ = the number of positive ions per cm

α = the degree of ionization

n = the number of molecules per cm^3

n_- = the number of negative ions per cm^3.

$$i = n_+ e \nu_- + n_-(-e)(-\nu_-) = ne(\nu_+ + \nu_-)\alpha$$

where i = the total current carried across a unit area

ν_+ and ν_- = the velocities with which the ions move through water

e = the electron charge.

$$\mu_+ = \frac{\nu_+}{E} \qquad \mu_- = \frac{\nu_-}{E}$$

where μ is the mobility of an ion, which is its velocity per unit field strength.

$$K = \frac{i}{E}$$

where K = conductivity

i = the current across a unit area

E = the electric field.

$$K = n_e(\mu_+ - \mu_-)\alpha$$

where $\quad \Lambda$ = the equivalent conductivity

N_0 = the Avogadro's number

N_{0_e} = the charge on a mole of electrons or the Faraday's constant, F.

$$\Lambda = F(\mu_+ - \mu_-)\alpha$$

$$\Lambda = z(\mu_+ - \mu_-)F$$

where z is the charge.

$$\alpha = \frac{\Lambda}{\Lambda_0}$$

$$\Lambda_0^+ = F\mu_+ \quad \text{and} \quad \Lambda_0^- = F\mu_-$$

For the reaction

$$AB \rightleftharpoons A^+ + B^-$$

$$c(1 - \alpha) \quad c\alpha \quad c\alpha$$

$$K = \frac{\alpha^2 c}{1 - \alpha}$$

where K is the equilibrium constant.

$$K = \left\{ \frac{(\Lambda/\Lambda_0)^2}{(1 - \frac{\Lambda}{\Lambda_0})} \right\} \quad C = \frac{\Lambda^2 C}{\Lambda_0(\Lambda_0 - \Lambda)}$$

The above equation is the Ostwald's Dilution Law.

11.2 COLLIGATIVE PROPERTIES OF IONIC SOLUTIONS

The colligative properties for nonelectrolytic solutions are less than that for electrolytic solutions.

The osmotic pressure of electrolytic solutions is given by the equation

$$\pi = icRT$$

where i is the Van't Hoff factor. The Van't Hoff factor is approximately equal to the number of ions formed in solution for strong electrolytes and is less than the number of ions but greater than unity for weak electrolytes.

The freezing-point depression for electrolytic solution is given by the equation

$$\Delta T_{fP} = imK_{fP}$$

where 'im' is the apparent molality.

The value of i, for weak electrolytes, is in a relationship with the degree of ionization, α by the equation

$$\alpha = \frac{i - 1}{\nu - 1}$$

where ν is the number of ions produced by dissociation.

11.3 TRANSFERENCE NUMBERS OF IONS

The transference number, t_i, or the transport number is the fraction of the total current carried by an ion.

$$t_+ + t_- = 1$$

The Hittorf method is used to determine t_i. In this method, after the current passes, the cell is divided into three sections which are analyzed for electrolyte content.

$$t_i = \frac{|N_o - N_f \pm Ne'|}{Ne}$$

where
t_i = the transference number

N_f = the final number of equivalents present

N_o = the original number of equivalents

Ne' = the number of equivalents involved in the electrode reaction (the positive sign is used if the equivalents are generated and the negative sign is used if they are removed).

Ne = the number of equivalents passed through the cell.

The value of Ne' is equal to either O or Ne, and this depends on the presence or absence of inert electrodes.

The moving boundary method is also used to determine t_i

$$t_i = \frac{F\ 1000c_i}{I}\ \frac{dV}{dt}$$

where
c_i = the concentration of the ion in equivalents per dm^3

I = the current in amperes

t = the time in seconds

V = the volume through which the moving boundary passes in m^3.

$$\mu_i = \frac{L}{t\left(\dfrac{dE}{dL}\right)}$$

where
μ_i = the ionic mobility

L = the distance that the moving boundary moves in m

t = time in seconds

$\dfrac{dE}{dL}$ = the electric field strength.

$$\frac{dE}{dL} = \frac{I}{AK}$$

where
I = the current, amperes

A = the area

K = a constant.

The ionic mobility is related to the transference number in the following equation:

$$t_i = \frac{\mu_i}{\mu_+ + \mu_-}$$

$$\lambda_i = t_i \Lambda \quad \text{or} \quad t_i = \frac{\lambda_i}{\Lambda}$$

where λ_i = the ionic equivalent conductance

and

$$\Lambda = \lambda_+ + \lambda_-$$

CHAPTER 12

ACTIVITIES OF IONS

12.1 UNI-UNIVALENT AND MULTIVALENT ELECTROLYTES

UNI-UNIVALENT ELECTROLYTE

$$a_{\pm} = (a_+ a_-)^{\frac{1}{2}}$$

where $a_{\pm}$ is the mean activity

$$\gamma_+ = \frac{a_+}{m} \qquad \gamma_- = \frac{a_-}{m}$$

where γ_+ and γ_- are the activity coefficients for the ions.

$$\gamma_{\pm} = [(\gamma_+)(\gamma_-)]^{\frac{1}{2}} = \left[\left(\frac{a_+}{m}\right)\left(\frac{a_-}{m}\right)\right]^{\frac{1}{2}} = \frac{a_{\pm}}{m}$$

MULTIVALENT ELECTROLYTE

For the simplest asymmetrical electrolyte that is of the form A_2B,

$$a_{\pm} = [(a_+)^2 (a_-)]^{\frac{1}{3}}$$

For infinite dilution

$$\left. \begin{array}{l} a_+ \to m_+ = 2m \\ a_- \to m_- = m \end{array} \right\} \quad \text{as } m \to o$$

$$a_\pm = [(2m)^2(m)]^{\frac{1}{3}} = (4)^{\frac{1}{3}} m$$

$$\gamma_\pm = \frac{a_\pm}{(4)^{\frac{1}{3}} m}$$

$$a_\pm = \gamma_\pm m_\pm$$

where

$a_\pm$ = the mean activity

$\gamma_\pm$ = the mean activity coefficient

$c_\pm$ = the mean ionic molar concentration.

For electrolyte in the form $A_x B_y$, the following equations are applicable:

$$a_\pm = ((a_+)^x(a_-)^y)^{\frac{1}{n}}$$

$$\gamma_\pm = ((\gamma_+)^x(\gamma_-)^y)^{\frac{1}{n}}$$

$$m_\pm = ((m_+)^x(m_-)^y)^{\frac{1}{n}}$$

$$n = x + y$$

$$a_\pm = \gamma_x \chi \qquad a_\pm = \gamma_m m$$

where

χ = the mole fraction

m = the molality.

$$\gamma_x = \gamma_\pm \left(\frac{d(\text{solution}) - M(\text{solute})C + M(\text{solvent})}{d(\text{solvent})} C_v \right)$$

$$\gamma_m = \gamma_\pm \left(\frac{C}{d(\text{solvent})m \ 10^{-3}} \right)$$

where

d = density in kg/m^3

M = molecular weight in g/mol

94

$$C = \text{molarity in } mol/dm^3$$

12.2 IONIC STRENGTH

The electrostatic forces between ions are dependent upon the charges and the concentrations of the individual ions. The effects of ionic charges are described by the ionic strength equation.

$$I = \tfrac{1}{2} \Sigma \, m_i z_i^2$$

where
I = the ionic strength of a solution

Σ = the summation for all the ions in solution

m_i = the molality

z_i = the charge on each ion.

For a uni-univalent electrolyte, the ionic strength is equal to the molality. For AB the ionic strength is

$$I = \tfrac{1}{2}[m(1)^2 + m(1)^2] = m$$

For AB_2

$$I = \tfrac{1}{2} [m(2)^2 + 2m(1)^2] = 3m$$

For a saturated aqueous solution of AB in equilibrium with excess solid AB,

$$K = a_{\pm}^2 = m_{\pm}^2 \, \gamma_{\pm}^2$$

where K is the equilibrium constant.

$$\ln \gamma_{\pm} = \ln \sqrt{K} - \ln m_{\pm}$$

$$\left. \begin{array}{l} \gamma_{\pm} \to 1 \\[4pt] \ln \gamma_{\pm} \to 0 \\[4pt] \ln \sqrt{K} \to \ln m_{\pm} \end{array} \right\} \quad \text{as } I \to 0$$

The following graphical method can be used to measure the activity coefficients for strong electrolytes.

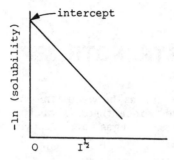

Fig. 12.1 The solubilty of a solid as a function of $(I)^{\frac{1}{2}}$ in solution.

$$\ln \gamma_\pm = \text{intercept} - \ln m$$

12.3 EXPERIMENTAL DETERMINATION OF THE ACTIVITY COEFFICIENT

For nonelectrolyte solutions, the osmotic coefficient ϕ is included in the equation

$$\ln a_A = \frac{-Mm}{1000} \phi$$

where a_A = the activity of the solvent

M = the molar mass of the solvent

m = the molal concentration.

For electrolyte solutions, the equation is

$$\ln a_A = -\frac{\nu Mm}{1000} \phi$$

where ν is the number of ions formed on dissociation. The Gibbs–Duhem equation is

$$d(\ln a_A) = -\frac{X_B}{X_A} d(\ln a_B)$$

$$d(\ln a_A) = \frac{-\nu mM}{1000} d(\ln a_{\pm})$$

$$d(\ln a_A) = \frac{-\nu mM}{1000} d(\ln(\gamma_{\pm}m_{\pm}))$$

$$d(\ln m_{\pm}) = d(\ln m)$$

$$d(\ln \gamma_{\pm}) = d\phi + (\phi - 1)d(\ln m)$$

$$\ln \gamma_{\pm} = \phi - 1 + \int_0^m (\phi - 1)d(\ln m)$$

$$\pi = \frac{RT}{\overline{V}_A} \frac{\nu M}{1000} m\phi$$

12.4 INTERIONIC ATTRACTIONS

$$\boxed{F = \frac{Q_1 Q_2}{kr^2}}$$

where F = the force between the two electric charges Q_1 and Q_2 that are separated by a distance r

k = the dielectric constant of the homogeneous medium that separates the two charges.

In Figure 12.2, the charge distribution in a solution is shown.

where d = the distance of closest approach of the ions and of the apparent diameter of the ions

ρ = the charge density in a differential spherical volume dv.

ze = the charge of ion

dv = $4\pi r^2 dr$ for a spherical shell.

$$\rho = \frac{Q}{dV}$$

where Q is the total charge in dV.

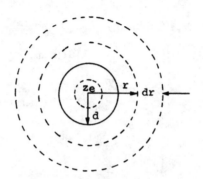

Fig. 12.2   The charge distribution
in a solution.

Poisson's equation of electrostatics is

$$\overline{V}^2 \psi = \frac{4\pi}{k} \rho$$

where ψ = the electric charge

$\overline{V}^2$ = the Laplacian operator and is equal to

$$\frac{\partial^2}{\partial x^2} + \frac{\partial^2}{\partial y^2} + \frac{\partial^2}{\partial z^2}$$

Poisson's equation, which is a function of r is written
as:

$$\frac{1}{r^2} \frac{d}{dr} \left(r^2 \frac{d\psi}{dr} \right) = \frac{-4\pi}{k} \rho$$

$$E = z_i e \psi$$

where   E = the energy of any individual ion at a
potential ψ

z_{ie} = the charge on the ion.

$$N_i = N_i^o \exp\left(\frac{-z_i e \psi}{kT}\right)$$

where
N_i = the actual concentration at the potential ψ

N_i^o = the uniform concentration at zero potential or in the absence of electric fields

k = the Boltzman constant

T = temerature

$$\rho = \sum_i z_i e N_i \exp\left(\frac{-z_i e \psi}{kT}\right)$$

Another form of Poisson's equation is

$$\frac{1}{r^2} \frac{d}{dr}\left(r^2 \frac{d\psi}{dr}\right) = b^2 \psi$$

where
$$b^2 = \frac{4\pi e^2}{KkT} \sum_i N_i z_i^2$$

$$\psi = \frac{A}{r} \exp(-br) \quad \text{only if } r > d \text{ (the apparent diameter of the ions)}$$

$$E = -\nabla \psi = -\frac{d\psi}{dr} = \frac{A(1 + br)}{r^2} \exp(-br)$$

$$E = -\frac{d\psi}{dr} = \frac{z_i e}{kr^2} \qquad r < d$$

$$A_i = \frac{Z_i e}{k}\left[\frac{\exp(bd)}{1 + bd}\right] \quad r = d$$

$$\psi(r) = \frac{Z_i e}{k}\left[\frac{\exp(bd)}{1 + bd}\right] \frac{\exp(-br)}{r}$$

where
$$b = \frac{2\pi^{\frac{1}{2}} e}{(KkT)^{\frac{1}{2}}} \sum_i (N_i Z_i^2)^{\frac{1}{2}}$$

99

12.5 THE DEBYE-HUCKEL LIMITING LAW

$$W_{el} = \int_{O}^{Z_e} \psi \, dQ$$

where Z_e is the total charge or the final charge and W_{el} is the electrical work that is associated with building up a charge dQ to a potential ψ.

The free energy for the overall process which involves transferring an ion from a solution of N_1 ions per cubic centimeter to a solution of N_2 ions per cubic centimeter is given by the equation,

$$\Delta G = RT \ln \frac{N_2}{N_1} + RT \ln \frac{\gamma_2}{\gamma_1}$$

At the central ion where $r = d$

$$\psi(d) = \frac{Ze}{kd(1 + bd)}$$

$$W_{el} = \frac{N_o Z^2 e^2}{2kd} \left\{ \frac{1}{1 + b_1 d} - \frac{1}{1 + b_2 d} \right\} = RT \ln \frac{\gamma_1}{\gamma_2}$$

where N_o is the Avogadro's number.

$$\ln \gamma_{\pm} = - |Z_+ Z_-| \frac{e^2}{2kRT} \left(\frac{b}{1 + bd} \right)$$

In the limit of infinite dilution

$$\ln \gamma_{\pm} = -1.172 |Z_+ Z_-| I^{\frac{1}{2}}$$

$$\log_{10} \gamma_{\pm} = -.509 |Z_+ Z_-| I^{\frac{1}{2}}$$

where I is the ionic strength.

$$\log_{10} \gamma_{\pm} = \frac{-.509 |Z_+ Z_-| I^{\frac{1}{2}}}{1 + BI^{\frac{1}{2}}}$$

where B is a constant dependent upon the distance of the closest approach of the ions.

CHAPTER 13

ELECTROCHEMICAL CELLS

13.1 VARIOUS TYPES OF CHEMICAL CELLS

In a galvanic cell, the anode is the electrode where oxidation occurs. The anode is a negatively charged electrode.

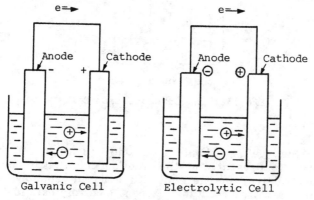

Fig. 13.1 A sample diagram for Galvanic and Electrolytic Cells.

The electrons move from the anode to the cathode. The cathode is the electrode where reduction occurs and is positively charged with respect to the anode. The cations which are positively charged move towards the cathode to react with the incoming electrons.

In an electrolytic cell, the anode is forced to be positively charged with respect to the cathode. The negatively charged anions move toward the anode to be oxidized. The positively charged cations move towards the cathode which is the negatively charged electrode to be reduced.

$$A \mid A^{2+} \mid B^{2+} \mid B$$

The vertical lines shown above, represent boundaries between phases.

In the concentration cell shown in Figure 13.2, ions pass across a liquid junction between two different solutions by transference. A junction potential is established across the boundary because the ions diffuse at different rates.

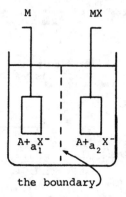

Fig.13.2 The Concentration Cell

In the cell without a liquid jucntion, a common electrolyte is present and the solution potential is common to both electrolytes. In the cell with a liquid junction, the two electrodes are immersed in different electrolytes and there is a potential difference across the interface of the two electrolytes.

In the concentration cell, the electrodes differ only in the amount of the electrolytes present per unit volume.

A salt bridge is used to eliminate the junction potential by connecting the bridge to the two half-cells.

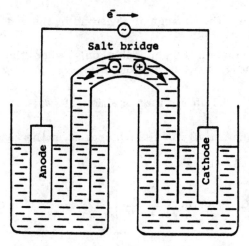

Fig. 13.3 A cell with a salt bridge.

13.2 EMF AND ELECTRODE POTENTIALS

$$\boxed{W_{el} = Q\xi}$$

where

W_{el} = electrical work

Q = charge

ξ = potential difference

$\Delta G = -W_{el}$ for a reversible work.

$\Delta G = -Q\xi$

$Q = nF$

$$\boxed{\Delta G = -nF\xi}$$

where

F = the faraday or the charge of a mole of electrons

$$n = \text{the number of electrons in the redox reaction.}$$

The above equation is applicable only when the reaction is reversible and ξ is the reversible emf.

$$\mu - \mu_i^\circ = RT \ln a_i \quad \text{for one mole of gas.}$$

where a_i is the activity of the ith species in the reaction

$$aA + bB = cC + dD$$

$$\Delta G = \Delta G^\circ + RT \ln \left(\frac{a_C^c \, a_D^d}{a_A^a \, a_B^b} \right) = \Delta G^\circ + RT \ln Q_a$$

$$\xi = \xi^\circ - \frac{RT}{nF} \ln Q_a$$

The above equation is the Nernst equation. $a_i = 1$ and $\xi = \xi^\circ$, when all the species are in their standard state.

ΔG and ξ are zeros at equilibrium

$$\xi^\circ = \frac{RT}{nF} \ln k_a \quad \text{at equilibrium}$$

$$\xi = \xi^\circ - \frac{.0592}{n} \log_{10}(Q) \quad \text{at } 298^\circ K$$

13.3 HALF-CELLS AND ELECTRODE POTENTIALS

The voltage of an electrochemical cell is the potential difference between the electrodes of each of the half-cells.

$$\xi = \xi_R - \xi_L$$

where $\quad \xi = \text{the cell emf}$

$\xi_R = \text{the emf of the right half-cell}$

ξ_L = the emf of the left half-cell.

When $\xi > 0$, reduction reaction occurs at the right half-cell or right-hand electrode.

$$M^+ + e^- \rightarrow M$$

At the left half-cell or left-hand electrode, oxidation tends to occur

$$M \rightarrow M^+ + e^-$$

Positive electrodes favor reduction and negative electrodes favor oxidation. Systems with high electrode potentials are reduced by systems with low electrode potentials.

$$\xi = \xi^\circ - \frac{RT}{nF} \ln Q_a$$

$$Q_a = Q_m Q_\gamma$$

$$\xi = \xi^\circ - \frac{RT}{nF} \ln Q_m - \frac{RT}{nF} \ln Q_\gamma$$

$$\xi + \frac{RT}{F} \ln m^2 = \xi^\circ - \frac{RT}{F} \ln \gamma_\pm^2$$

$$\xi + \frac{2RT}{F} \ln m = \xi^\circ - \frac{2RT}{F} \ln \gamma_\pm$$

Using the Debye–Huckel theory

$$\ln \gamma_\pm = B m^{\frac{1}{2}}$$

$$\xi + \frac{2RT}{F} \ln m = \xi^\circ - \frac{2RTB}{F} m^{\frac{1}{2}}$$

where B is a constant.

The standard potential of the cell is determined by the graph shown in Figure 13.4.

The values of the standard electrode potentials in aqueous solutions at 25°C are listed below in Table 13.1.

Table 13.1

Standard electrode potentials in aqueous solutions at 25°C.

Electrode	Electrode reaction	E°(V)
ACID SOLUTIONS		
$Li\|Li^+$	$Li^+ + e^- \rightleftharpoons Li$	-3.045
$K\|K^+$	$K^+ + e^- \rightleftharpoons K$	-2.925
$Ba\|Ba^{2+}$	$Ba^{2+} + 2e^- \rightleftharpoons Ba$	-2.906
$Ca\|Ca^{2+}$	$Ca^{2+} + 2e^- \rightleftharpoons Ca$	-2.87
$Na\|Na^+$	$Na^+ + e^- \rightleftharpoons Na$	-2.714
$La\|La^{3+}$	$La^{3+} + 3e^- \rightleftharpoons La$	-2.52
$Mg\|Mg^{2+}$	$Mg^{2+} + 2e^- \rightleftharpoons Mg$	-2.363
$Th\|Th^{4+}$	$Th^{4+} + 4e^- \rightleftharpoons Th$	-1.90
$U\|U^{3+}$	$U^{3+} + 3e^- \rightleftharpoons U$	-1.80
$Al\|Al^{3+}$	$Al^{3+} + 3e^- \rightleftharpoons Al$	-1.66
$Mn\|Mn^{2+}$	$Mn^{2+} + 2e^- \rightleftharpoons Mn$	-1.180
$V\|V^{2+}$	$V^{2+} + 2e^- \rightleftharpoons V$	-1.18
$Zn\|Zn^{2+}$	$Zn^{2+} + 2e^- \rightleftharpoons Zn$	-0.763
$Tl\|TlI\|I^-$	$TlI(s) + e^- \rightleftharpoons Tl + I^-$	-0.753
$Cr\|Cr^{3+}$	$Cr^{3+} + 3e^- \rightleftharpoons Cr$	-0.744
$Tl\|TlBr\|Br^-$	$TlBr(s) + e^- \rightleftharpoons Tl + Br^-$	-0.658
$Pt\|U^{3+},U^{4+}$	$U^{4+} + e^- \rightleftharpoons U^{3+}$	-0.61
$Fe\|Fe^{2+}$	$Fe^{2+} + 2e^- \rightleftharpoons Fe$	-0.440
$Cd\|Cd^{2+}$	$Cd^{2+} + 2e^- \rightleftharpoons Cd$	-0.403
$Pb\|PbSO_4\|SO_4^{2-}$	$PbSO_4 + 2e^- \rightleftharpoons Pb + SO_4^{2-}$	-0.359
$Tl\|Tl^+$	$Tl^+ + e^- \rightleftharpoons Tl$	-0.3363
$Ag\|AgI\|I^-$	$AgI + e^- \rightleftharpoons Ag + I^-$	-0.152
$Pb\|Pb^{2+}$	$Pb^{2+} + 2e^- \rightleftharpoons Pb$	-0.126
$Pt\|D_2\|D^+$	$2D^+ + 2e^- \rightleftharpoons D_2$	-0.0034
$Pt\|H_2\|H^+$	$2H^+ + 2e^- \rightleftharpoons H_2$	0.0000
$Ag\|AgBr\|Br^-$	$AgBr + e^- \rightleftharpoons Ag + Br^-$	+0.071
$Ag\|AgCl\|Cl^-$	$AgCl + e^- \rightleftharpoons Ag + Cl^-$	+0.2225
$Pt\|Hg\|Hg_2Cl_2\|Cl^-$	$Hg_2Cl_2 + 2e^- \rightleftharpoons 2Cl^- + 2Hg(l)$	+0.2676
$Cu\|Cu^{2+}$	$Cu^{2+} + 2e^- \rightleftharpoons Cu$	+0.337
$Pt\|I_2\|I^-$	$I_3^- + 2e^- \rightleftharpoons 3I^-$	+0.536
$Pt\|O_2\|H_2O_2$	$O_2 + 2H^+ + 2e- \rightleftharpoons H_2O_2$	+0.682
$Pt\|Fe^{2+},Fe^{3+}$	$Fe^{3+} + e^- \rightleftharpoons Fe^{2+}$	+0.771
$Ag\|Ag^+$	$Ag^+ + e^- \rightleftharpoons Ag$	+0.7991
$Au\|AuCl_4^-,Cl^-$	$AuCl_4^- + 3e^- \rightleftharpoons Au + 4Cl^-$	+1.00
$Pt\|Br_2\|Br^-$	$Br_2 + 2e^- \rightleftharpoons 2Br^-$	+1.065
$Pt\|Tl^+,Tl^{3+}$	$Tl^{3+} + 2e^- \rightleftharpoons Tl^+$	+1.25
$Pt\|H^+,Cr_2O_7^{2-},Cr^{3+}$	$Cr_2O_7^{2-} + 14H^+ + 6e^- \rightleftharpoons 2Cr^{3+} + 7H_2O$	+1.33
$Pt\|Cl_2\|Cl^-$	$Cl_2 + 2e^- \rightleftharpoons 2Cl^-$	+1.3595
$Pt\|Ce^{4+},Ce^{3+}$	$Ce^{4+} + e^- \rightleftharpoons Ce^{3+}$	+1.45
$Au\|Au^{3+}$	$Au^{3+} + 3e^- \rightleftharpoons Au$	+1.50
$Pt\|Mn^{2+},MnO_4^-$	$MnO_4^- + 8H^+ + 5e^- \rightleftharpoons Mn^{2+} + 4H_2O$	+1.51
$Au\|Au^+$	$Au^+ + e^- \rightleftharpoons Au$	+1.68
$PbSO_4\|PbO_2\|H_2SO_4$	$PbO_2 + SO_4 + 4H^+ + 2e^- \rightleftharpoons PbSO_4 + 2H_2O$	+1.685
$Pt\|F_2\|F^-$	$F_2(g) + 2e^- \rightleftharpoons 2F^-$	+2.87
BASIC SOLUTIONS		
$Pt\|SO_3^{2-},SO_4^{2-}$	$SO_4^{2-} + H_2O + 2e^- \rightleftharpoons SO_3^{2-} + 2OH^-$	-0.93
$Pt\|H_2\|OH^-$	$2H_2O + 2e^- \rightleftharpoons H_2 + 2OH^-$	-0.828
$Ag\|Ag(NH_3)_2^+,NH_3(aq)$	$Ag(NH_3)_2^+ + e^- \rightleftharpoons Ag + 2NH_3 \ (aq)$	+0.373
$Pt\|O_2\|OH^-$	$O_2 + 2H_2O + 4e^- \rightleftharpoons 4OH^-$	+0.401
$Pt\|MnO_2\|MnO_4^-$	$MnO_4^- + 2H_2O + 3e^- \rightleftharpoons MnO_2 + 4OH^-$	+0.588

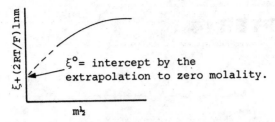

Fig. 13.4 A graphical method to determine
the standard potential.

13.4 THERMODYNAMIC DATA FROM CELL EMFS

$$\Delta S = -\left(\frac{\partial \Delta G}{\partial T}\right)_P = nF\left(\frac{\partial \xi}{\partial T}\right)_P$$

where $n = 1$.

$$\Delta H = \Delta G + T\Delta S$$

$$\Delta H = -nF\xi + nFT\left(\frac{\partial \xi}{\partial T}\right)_P$$

$$\Delta C_p = nFT\left(\frac{\partial^2 \xi}{\partial T^2}\right)_P$$

$$\Delta V = -nF\left(\frac{\partial \xi}{\partial P}\right)_T$$

CHAPTER 14

PHENOMENOLOGICAL RATES OF CHEMICAL REACTIONS

14.1 RATE OF A SIMPLE PROCESS AND ITS HALF-LIFE

$$-\frac{dc}{dt} = kc$$

where c = the concentration of molecules

            k = a constant

The half-life, $t\frac{1}{2}$, is the time a process takes to be half-completed. When $t - t\frac{1}{2}$, $c = \frac{1}{2}c_0$, where c_0 is the initial concentration.

$$\log t\tfrac{1}{2} = \log\left[\frac{2^{n-1} - 1}{(n-1)k}\right] - (n - 1)\log c_0$$

where $t\frac{1}{2}$ = half-life period

            c_0 = initial concentration

            $n-1$ = slope of the plot of $\log t\frac{1}{2}$ against $\log c_0$.

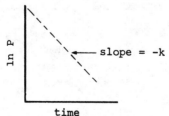

Fig. 14.1   ln P as a function of time.

108

$$-\frac{dp}{dt} = kP$$

$$P = P_0 \exp(-kt)$$

$$P = \tfrac{1}{2}P_0 = P_0 \exp(-kt\tfrac{1}{2})$$

$$t\tfrac{1}{2} = \frac{.693}{k}$$

14.2 EMPIRICAL ORDER OF A CHEMICAL REACTION

The rate of reaction is the rate of change of the concentration of any reactant or any product.

For the reaction,

$$aA + bB + cC + \ldots \rightarrow dD + eE + fF + \ldots$$

$$R = -\frac{d[A]}{dt} = -\frac{a}{b}\frac{d[B]}{dt} = -\frac{a}{c}\frac{d[C]}{dt} = \frac{d}{a}\frac{d[D]}{dt} = \frac{e}{a}\frac{d[E]}{dt}$$

$$R = -\frac{d[A]}{dt} = -\frac{dC_A}{dt} = f(C_A, C_B, \ldots)$$

$$-\frac{dC_A}{dt} = k\, C_A^{\alpha}\, C_B^{\beta}\, C_C^{\gamma}$$

where $\quad C_i$ = the concentration of all species

$\qquad k$ = the rate constant

α, β, γ = the partial order of the reaction

$\qquad \alpha$ = the partial order of the reaction with respect to A

$\alpha + \beta + \gamma$ = the overall order of the reaction or the order

$$\alpha + \beta + \gamma = \zeta$$

The units of k are $mol^{-(\zeta-1)}$ $liter^{(\zeta-1)}S^{-1}$ when the concentration units are moles per liter.

For the reaction,
$$2A + B_2 \rightarrow 2AB$$

$$\frac{-d[A]}{dt} = \frac{-2d[B_2]}{dt} = \frac{d[AB]}{dt}$$

$$\frac{-d[A]}{dt} = k_A[A]^2[B_2]$$

$$\frac{-d[B_2]}{dt} = k_{B_2}[A]^2[B_2]$$

$$\frac{d[AB]}{dt} = k_{AB}[A]^2[B_2]$$

$$k_A = 2k_{B_2} = k_{AB}$$

ZERO-ORDER REACTIONS

Zero-order reactions have the following form of differential equation:

$$\frac{-dC}{dt} = k$$

$$C = C_0 - kt$$

$$C = C_0 \quad \text{when } t = 0.$$

k is equal to the slope of the line obtained by plotting c as a function of t and C_0 is the intercept of the line.

14.3 FIRST-ORDER REACTIONS

The first-order reaction has the form A $\rightarrow$ products and the rate equation is

$$\text{rate} = \frac{-dC}{dt} = kC$$

when $t = 0$, $C = C_0$

$$\ln \frac{C}{C_0} = -kt \qquad \frac{C}{C_0} = e^{-kt}$$

$$\ln C = \ln C_0 - kt$$

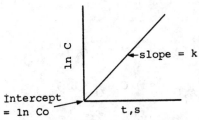

Fig. 14.2 A graphical method to determine the rate constant and the initial concentration.

$$P = cRT$$

$$c = P/RT$$

$$\frac{C}{C_0} = \frac{P/RT}{P_0/RT} = e^{-kt}$$

14.4 SECOND-ORDER REACTIONS

$$aA + bB \rightarrow products$$

If $a = b$, and $C_{A,O} \neq C_{B,O}$, then the rate equation is

$$\boxed{\frac{-dC_A}{dt} = kC_A C_B = k(C_{A,O} - x)(C_{B,O} - x)}$$

where x is the number of moles per liter of A that have reacted at time, t.

$$\frac{dx}{dt} = k(C_{A,O} - x)(C_{B,O} - x)$$

$$kt = \int \frac{dx}{(C_{A,O} - x)(C_{B,O} - x)}$$

111

$$kt = \left\{ \frac{1}{C_{A,O} - C_{B,O}} \right\} \ln \left\{ \frac{C_{B,O} C_A}{C_{A,O} C_B} \right\}$$

If $a \neq b$, and $C_{A,O} \neq C_{b,O}$, then the rate equation is

$$\frac{-dC_A}{dt} = k(C_{A,O} - ax)(C_{B,O} - bx)$$

$$\frac{dx}{dt} = k(C_{A,O} - ax)(C_{B,O} - bx)$$

$$kt = \left\{ \frac{1}{bC_{A,O} - aC_{B,O}} \right\} \ln \left\{ \frac{C_{B,O} C_A}{C_{A,O} C_B} \right\}$$

The rate law for the second order reaction $2A \rightarrow$ Products is

$$\frac{-dC}{dt} = kC^2$$

$$\frac{1}{C} = \frac{1}{C_0} + kt$$

k has the units of $(time)^{-1} (concentration)^{-1}$.

$$t\tfrac{1}{2} = (kC_0)^{-1}$$

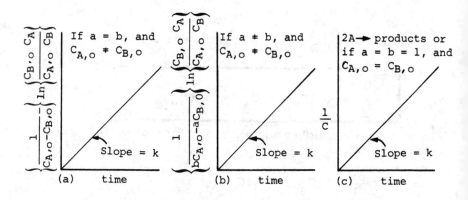

Fig. 14.3 Graphical method to determine the rate constant for the different forms of second order reactions.

112

14.5 THIRD-ORDER REACTIONS

The differential rate equation for the reaction

$$A + B + C \rightarrow \text{products}$$

where $C_{A,O} \neq C_{B,O} \neq C_{C,O}$ is

$$\frac{\ln(C_A/C_{A,O})}{(C_{A,O}-C_{B,O})(C_{C,O}-C_{A,O})} + \frac{\ln(C_B/C_{B,O})}{(C_{A,O}-C_{B,O})(C_{B,O}-C_{C,O})}$$

$$+ \frac{\ln(C_C/C_{C,O})}{(C_{B,O}-C_{C,O})(C_{C,O}-C_{A,O})} = kt$$

For the reaction $A + B + C \rightarrow$ Products, with $C_{B,O} \neq C_{A,O} = C_{C,O}$ or for the reaction $2A + B \rightarrow$ products, where $C_{A,O} \neq C_{B,O}$ or $C_{A,O} \neq 2C_{B,O}$, the rate equation is

$$\frac{-dC_A}{dt} = kC_A^2 C_B$$

$$\frac{2}{(2C_{B,O}-C_{A,O})^2}\left[\frac{2(2C_{B,O}-C_{A,O})(C_{A,O}-C_A)}{C_{A,O}C_A} + \ln\frac{C_{B,O}C_A}{C_{A,O}C_B}\right]$$

$$= kt$$

For the reaction $A + B \rightarrow$ products, the integrated rate equation is

$$\frac{1}{(C_{B,O}-C_{A,O})^2}\left[\frac{(C_{B,O}-C_{A,O})(C_{A,O}-C_A)}{C_{A,O}C_A} + \ln\frac{C_{B,O}C_A}{C_{A,O}C_B}\right] = kt$$

For the reaction $A + B + C \rightarrow$ Products, with $C_{A,O} = C_{B,O} = C_{C,O}$, or for the reaction $2A + B \rightarrow$ Products, with $C_{A,O} = C_{B,O}$ or $C_{A,O} = 2C_{B,O}$, or for the reaction $3A \rightarrow$ Products, the differential rate equation is

$$\frac{-dC}{dt} = kC^3$$

$$\frac{1}{2}\left(\frac{1}{C^2} - \frac{1}{C_0^2}\right) = kt$$

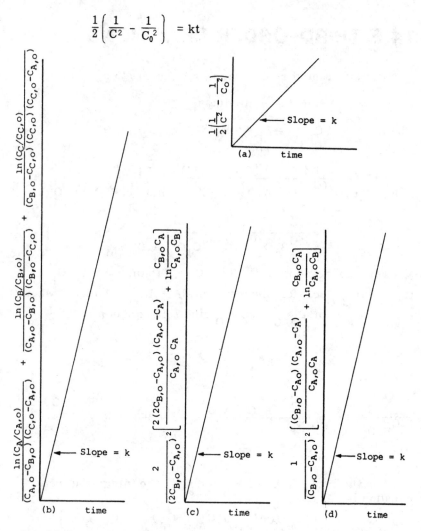

Fig. 14.4 Graphical method to determine the rate constant.

14.6 DETERMINING THE ORDER OF A REACTION

The following methods are useful to determine the order and the rate constant of a reaction.

1) The half-life method

$$\ln t_{\frac{1}{2}} = \ln \frac{2^{n-1} - 1}{(n-1)k} - (n-1)\ln C_0$$

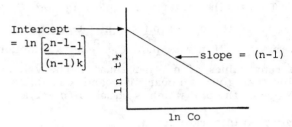

Intercept $= \ln \left[\dfrac{2^{n-1}-1}{(n-1)k}\right]$

slope $= (n-1)$

$\ln t_{\frac{1}{2}}$

$\ln C_0$

Fig. 14.5   The graphical half-life method.

2) The differential method

$$\text{rate} = \frac{\pm dC_i}{dt} = kC_i^{\,n}$$

where C_i is the concentration of a substance while all the other substances are present in excess or fixed amount.

$$\ln \text{rate} = \ln \left(\pm \frac{dC_i}{dt} \right) = \ln k + n \ln C_i$$

$\left(\dfrac{\pm dC_i}{dt}\right)$ can be determined from the tangent of the curve obtained by plotting C_i as a function of time.

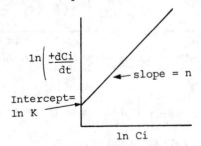

$\ln \left(\dfrac{\pm dC_i}{dt} \right)$

slope $= n$

Intercept $= \ln K$

$\ln C_i$

Fig. 14.6 The graphical differential method.

3. The Integral Methods

A) The graphical integral method

The graphical integral method is used to determine n and k by trial and error in the equation $\ln\left(\dfrac{\pm dC_i}{dt}\right) = \ln k + n \ln C_i$. This method starts by plotting $\log C_i$ as a function of t. If the plot is linear, $n = 1$ and if the plot is a curve $n \neq 1$. If $n \neq 1$, plot C_i^{1-n} as a function of t for different values of n other than 1. The value of n is the value that gives a linear plot, and k is determined from the slope of the plot which is equal to $(n - 1)k$.

B) The mathematical integral method

This method is used to determine k from the integrated form of the rate equation by trial and error. The value of k is determined from the integrated rate equation for different values of n. The value of n is the value that gives the same rate constant for all the data.

4) The method of initial rates

The rate law for a reaction between A and B is

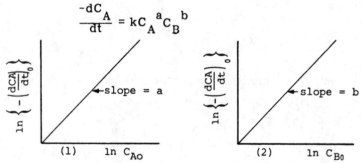

14.7 The graphical initial rates method used to determine the orders a and b in figures (1) and (2) respectively.

$$\ln\left\{-\left(\frac{dC_A}{dt}\right)_0\right\} = \ln k + a \ln C_{A_0} + b \ln C_{B_0}$$

at $t = 0$, the rate equation is

$$\left\{-\left(\frac{dC_A}{dt}\right)_0\right\} = k C_{A_0}^{a} C_{B_0}^{b}$$

116

5) The isolation method

When a reactant is present in large excess, its concentration is considered to be constant. For the reaction $A + B + C \rightarrow$ products, the rate equation is

$$\frac{-dC_A}{dt} = kC_A{}^a C_B{}^b C_C{}^c$$

If the value of A is much smaller than that of B and C, then the concentrations of B and C are constant and the rate equation would change to

$$\frac{-dC_A}{dt} = k'C_A{}^a$$

where $k' = kC_B{}^b C_C{}^c$.

The rate law $\frac{-dC_A}{dt} = k'C_A{}^a$ is the pseudo-first-order rate law of the third order rate law.

6) Relaxation methods

The relaxation methods are useful for the study of fast reactions.

$$\Delta C_i = \Delta C_{i,o} e^{-t/\tau}$$

where

ΔC_i = the displacement from equilibrium at time t

$\Delta C_{i,o}$ = the initial displacement from equilibrium

τ = the relaxation time.

$$\frac{d(\Delta C_i)}{dt} = -\Delta C_i\left(\frac{1}{\tau}\right)$$

THE PERIODIC TABLE

METALS — **NONMETALS**

TRANSITION METALS

KEY:
- 112.40 — Atomic weight
- **Cd** — Symbol
- 48 — Atomic number

PERIODS	IA	IIA	IIIB	IVB	VB	VIB	VIIB	VIII	VIII	VIII	IB	IIB	IIIA	IVA	VA	VIA	VIIA	O
1	1.0079 **H** 1																	4.00260 **He** 2
2	6.94 **Li** 3	9.01218 **Be** 4											10.81 **B** 5	12.011 **C** 6	14.0067 **N** 7	15.9994 **O** 8	18.9984 **F** 9	20.179 **Ne** 10
3	22.9898 **Na** 11	24.305 **Mg** 12											26.9815 **Al** 13	28.086 **Si** 14	30.9738 **P** 15	32.06 **S** 16	35.453 **Cl** 17	39.948 **Ar** 18
4	39.098 **K** 19	40.08 **Ca** 20	44.9559 **Sc** 21	47.90 **Ti** 22	50.9414 **V** 23	51.996 **Cr** 24	54.9380 **Mn** 25	55.847 **Fe** 26	58.9332 **Co** 27	58.71 **Ni** 28	63.546 **Cu** 29	65.38 **Zn** 30	69.72 **Ga** 31	72.59 **Ge** 32	74.9216 **As** 33	78.96 **Se** 34	79.904 **Br** 35	83.80 **Kr** 36
5	85.4678 **Rb** 37	87.62 **Sr** 38	88.9059 **Y** 39	91.22 **Zr** 40	92.9064 **Nb** 41	95.94 **Mo** 42	98.9062 **Tc** 43	101.07 **Ru** 44	102.9055 **Rh** 45	106.4 **Pd** 46	107.868 **Ag** 47	112.40 **Cd** 48	114.82 **In** 49	118.69 **Sn** 50	121.75 **Sb** 51	127.60 **Te** 52	126.9046 **I** 53	131.30 **Xe** 54
6	132.9054 **Cs** 55	137.34 **Ba** 56	57–71 *	178.49 **Hf** 72	180.9479 **Ta** 73	183.85 **W** 74	186.2 **Re** 75	190.2 **Os** 76	192.22 **Ir** 77	195.09 **Pt** 78	196.9665 **Au** 79	200.59 **Hg** 80	204.37 **Tl** 81	207.2 **Pb** 82	208.9804 **Bi** 83	(210) **Po** 84	(210) **At** 85	(222) **Rn** 86
7	(223) **Fr** 87	(226.0254) **Ra** 88	89–103 †	(260) **Ku** 104	(260) **Ha** 105													

* LANTHANIDE SERIES

138.9055 **La** 57	140.12 **Ce** 58	140.9077 **Pr** 59	144.24 **Nd** 60	(145) **Pm** 61	150.4 **Sm** 62	151.96 **Eu** 63	157.25 **Gd** 64	158.9254 **Tb** 65	162.50 **Dy** 66	164.9304 **Ho** 67	167.26 **Er** 68	168.9342 **Tm** 69	173.04 **Yb** 70	174.97 **Lu** 71

† ACTINIDE SERIES

(227) **Ac** 89	232.0381 **Th** 90	231.0359 **Pa** 91	238.029 **U** 92	237.0482 **Np** 93	(242) **Pu** 94	(243) **Am** 95	(245) **Cm** 96	(245) **Bk** 97	(248) **Cf** 98	(253) **Es** 99	(254) **Fm** 100	(256) **Md** 101	(253) **No** 102	(257) **Lr** 103

REA's **Problem Solvers**

The "PROBLEM SOLVERS" are comprehensive supplemental textbooks designed to save time in finding solutions to problems. Each "PROBLEM SOLVER" is the first of its kind ever produced in its field. It is the product of a massive effort to illustrate almost any imaginable problem in exceptional depth, detail, and clarity. Each problem is worked out in detail with a step-by-step solution, and the problems are arranged in order of complexity from elementary to advanced. Each book is fully indexed for locating problems rapidly.

*If you would like more information about any of these books,
complete the coupon below and return it to us or visit your local bookstore.*
